CARTOGRAPHIC MEXICO

A BOOK IN THE SERIES

LATIN AMERICA OTHERWISE:

Languages, Empires, Nations

Series editors:

Walter D. Mignolo, DUKE UNIVERSITY

Irene Silverblatt, DUKE UNIVERSITY

Sonia Saldívar-Hull, UNIVERSITY OF

CALIFORNIA, LOS ANGELES

Maps might appear to be objective records of an objective geography, but, as Raymond Craib has shown, they are anything but. In his sweeping study of nineteenth- and twentieth-century maps drawn under the auspices of the Mexican government, Craib presents the political side of maps—an underside often hidden by "mapmaking's" technocratic surface. For these maps didn't simply reflect a spatial reality, they created one. Maps were politics: they made Mexico appear to be a unified, coherent, entity, and they did so, in part, by plastering over conflicts between government agencies and local communities. Maps were doubly functional: they gave Mexico internal definition and were a key to its membership in the international club of nations.

With great originality and insight, Craib has inspected a "taken-for-granted" of modern life and revealed its immersion in relations of power. Nonetheless, he is careful to record the social contradictions and complications of mapmaking: while maps were a tool of power, they were also challenged; while maps expressed the will of government agencies, they were also lightening rods of resistance. Maps could never totally dominate the geographical vision of Mexico's subjects, and, to bear out Craib's argument, Mexico is still confounded by conflicts rooted in a question of maps, in questions of the state's version of territorial—and political—control. Craib, by penetrating the surface of something as ordinary as maps, has helped us see Latin America otherwise.

CARTOGRAPHIC
MEXICO

A HISTORY OF STATE FIXATIONS

AND FUGITIVE LANDSCAPES

Raymond B. Craib

Duke University Press

Durham and London

2004

© 2004 DUKE UNIVERSITY PRESS

All rights reserved

Printed in the United States

of America on acid-free paper ∞

Designed by Amy Ruth Buchanan

Typeset in Janson by Wilsted &

Taylor Publishing Services

Library of Congress Cataloging-in-

Publication Data appear on the last

printed page of this book.

Publication of this book was made possible
by a subvention granted by the Hull Memorial
Publication Fund of Cornell University.

An earlier version of chapter 1 was originally
published as "A Nationalist Metaphysics: State
Fixations, National Maps, and the Geo-Historical
Imagination in Nineteenth-Century Mexico,"
Hispanic American Historical Review 82: 1 (2002).
An earlier version of chapter 3 was originally
published as "Standard Plots and Rural Resistance,"
in *The Mexico Reader: History, Culture, Politics,*
edited by Gilbert M. Joseph and Timothy J.
Henderson. Durham: Duke University
Press, 2003.

FOR CYNTHIA

CONTENTS

ILLUSTRATIONS

✳
ABBREVIATIONS

documento	doc.
expediente	exp.
folio	f.
legajo	leg.
Mapoteca Manuel Orozco y Berra	MOB
Colección General—Veracruz	CGV
Colección General—República Mexicana	CGRM
Colección General—Comisión Geográfico-	
Exploradora	CGE
paquete	paq.
Sociedad Mexicana de Geografía y Estadística	SMGE

✳

ACKNOWLEDGMENTS

There are two acknowledgments that need to be made before all others. I have no idea where I would be now if it had not been for Dick Goff. To Dick this book is a meager offering for all his years of kindness, which began at Eastern Michigan University. I can never repay him. The University of New Mexico's Latin American Institute, a truly wonderful institution, took a chance and provided me the initial opportunity and funds to go to graduate school. Thank you especially to Theo Crevenna, Nita Daly, Eric García, Linda Hall, Robert Himmerich y Valencia, Sharon Kellum, Linda Kjeldgaard, Manya Paul, Enrique Semo, and Joanie Swanson who all made my time there a real pleasure.

Little gets done without funding and I am grateful to the following programs for supporting my research: Yale University's Program in Agrarian Studies; the Yale Council on International and Area Studies; a Social Science Research Council International Dissertation Field Research Grant with funds provided by the Andrew W. Mellon Foundation; a Fulbright-Hays Doctoral Dissertation Grant; an Albert J. Beveridge Grant from the American Historical Association; and a National Endowment for the Humanities Summer Stipend. A Whiting Fellowship in the Humanities and a Mellon fellowship from the Yale Council on Latin American Studies provided funds for the writing of the dissertation out of which this book grew. The Cornell University Department of History Faculty Research Fund generously covered the costs of photographing maps and images.

Thanks also to the staff and archivists at the Benson Library's Latin American Collection at the University of Texas at Austin, the Archivo Histórico del Agua, the Archivo General de la Nación, the Archivo

General Agrario, the Archivo Municipal de Orizaba, the Centro de Estudios de Historia de Mexico, and the Sociedad Mexicana de Geografía y Estadística. My lengthy stints at the Mapoteca Manuel Orozco y Berra in Mexico City and the Archivo General del Estado de Veracruz in Xalapa were productive and pleasurable thanks to the assistance of Victor Hernández Ortíz and Carlos Vidalí (at the Mapoteca) and Olivia Domínguez Pérez and her wonderful staff (at the AGEV). My work in the municipal archives of Acultzingo and Misantla was both possible and enjoyable due to the assistance of Julio Palacios Martínez (in Acultzingo) and Ignacio de León (in Misantla). Finally, I am greatly indebted to Carmen Boone de Aguilar, who welcomed me into her home, permitted me to consult the archives of her granduncle, Francisco Canóvas y Pasquel, and allowed me to reproduce a number of the photographs in this book.

Numerous people have shaped this work at various stages and in a variety of ways with their friendship, collegiality, and intellectual engagement. Thanks to Luis Aboites, Jonathan Amith, Steve Bachelor, Amy Chazkel, Carmen Blázquez Domínguez, Cynthia Brock, D. Graham Burnett, Karen Caplan, Matthew Edney, Antonio Escobar Ohmstede, Chris Gill, Todd Hartch, Karl Jacoby, Ben Johnson, Bernardo Michael, John Noyes, Jolie Olcott, Erika Pani, Mauricio Tenorio-Trillo, Michael Werner, and Eric Worby. I am particularly grateful to Rolena Adorno, John Mack Faragher, and Stuart Schwartz, all of whom read drafts of this project at various stages and offered extensive advice and commentary; to Emilia Viotti da Costa, although she may not wish to be implicated in any form with what follows; to Jonathan Spence for his support and a well-timed letter to Honolulu; and to Kay Mansfield at Yale's Program in Agrarian Studies for everything.

I was fortunate to have the company of Mike Ducey and Heather Fowler-Salamini while in Xalapa. They both patiently answered numerous questions and asked probing ones of their own that made me think about my project in different ways. Mike, carrying out his own research on land divisions in Veracruz, generously shared his findings and extensive knowledge. Much of chapters 2 and 3 bear the mark of his generosity. The people of El Grande, Veracruz, were especially gracious in answering numerous questions and allowing me to attend ejidatario meetings. My time in Mexico over the years has always been a real pleasure due to the hospitality of Carmen and Brendan Rowlands, Carmen

Piña, Luis Moreno, and Héctor Mendoza Vargas. This project never would have gotten off the ground in the first place if it had not been for Héctor. He took an immediate interest when we first met over coffee in (of course) Gandhi and has since been a persistent interlocutor, saving me from numerous errors and patiently explaining (yet again) the complexities of cartographic practice and theory. Any errors that remain are there in spite of his best efforts; much of what is good about this book is due in no small part to his help.

Many of the ideas in this book have been hashed out, argued over, and developed in conversations spanning many years with Rob Campbell, James Kessenides, Rick López, Héctor Mendoza Vargas, Mark Overmyer-Velázquez, and Daviken Studnicki-Gizbert. They were all writing their own works but still found time to read various chapters, offer sharp comments, suggest readings, and help me through various (and numerous) points of frustration. I want to recognize their contributions but especially their friendship. As I sought to turn the dissertation into a book, Peter Dear, Heather Fowler-Salamini, Emilio Kourí, Aldo Lauria-Santiago, Héctor Mendoza Vargas, and Mary Roldán kindly agreed to read various chapters, offered excellent feedback, and saved me from an array of errors, as did the two anonymous reviewers for Duke University Press. I have not taken all of their advice but I hope they see a reflection of their efforts here. Valerie Millholland's reputation led me to Duke University Press and I have not been disappointed: I doubt I could have found a more supportive, understanding, and attentive editor. My colleagues and the staff in the History Department at Cornell have created an extraordinarily collegial, warm environment within which to teach and write. I would like to thank in particular María Cristina García, Peter Holquist, Tamara Loos, Mary Roldán, and Eric Tagliacozzo for their encouragement, advice, and support.

In the course of writing the dissertation out of which this book grew, Jim Scott generously shared not only his knowledge and time but his space. Once my son was born, writing at home became much more difficult. Jim offered me squatting rights to his new office. It is a testament to Jim's generous spirit that this would come as no surprise to all who know him. Without the privacy and tranquility of that office (and the high-powered espresso machine down the hall) this book would have been much longer in the making. There is little I can say about Gil Joseph that most people in this field do not already know. He is a gem. He has been a passionate teacher, a tireless and selfless advisor, a model mentor, and a good friend. When I met with graduate students at a university where I was interviewing for a job, one of them asked me what things about my advisor I would emulate and what things I would do different. I had no answer to the second half of that question.

Finally comes my family. James Brock, as well as taking my money at mah-jongg, took an interest in this project and answered many of my questions regarding the art and practice of surveying (although I have a feeling he may not agree with much of what I have written). Over the course of researching and writing this book, my sister, Linda, has moved in and out of my life but she has *always* been an inspiring presence and been there when I needed her. Birthing has been a major part of my life in recent years. I would still be laboring with this book if it were not for the help and love of some amazing people: Mary Brock, who has babysat, cooked, brought home the vino, and done just about everything imaginable to make day-to-day life just a little less hectic; and my parents, Raymond and Julia Craib, who have been an endless source of encouragement, laughter, love, and emotional support. Thank you.

Regardless of the involvement and participation of so many people, researching and writing a book is a solitary, consuming process. Cynthia Brock, and our children Connor and Alana, brought me back to the world every day and put it all in perspective for me. Connor and Alana have been living with this book since they were born. They did not slow it down one bit; on the contrary, I knew the sooner I finished, the sooner I could get back to playing chase and hide-and-seek, swinging in the hammock, splashing in puddles, and watching two wonderful children grow. Cynthia Brock has put up with constant moves, fieldwork separations, my lame contributions to our bank account, and my neuroses. She has never complained and she has made so much possible: this is for her.

＊

Fixations constitute procedures for forgetting. The trace left behind is substituted for practice. It exhibits the (voracious) property that the geographical system has of being able to transform action into legibility, but in doing so it causes a way of being in the world to be forgotten.

— MICHEL DE CERTEAU, *The Practice of Everyday Life*

INTRODUCTION

Writing a Spatial History of Modern Mexico

> Geography is not an immutable thing. It is made, it is remade
> everyday; at each instant, it is modified by men's actions.
> —Elisée Reclus, *L'Homme et la terre*

In 1985, on the cusp of signing on to the GATT (General Agreement on Tariffs and Trade), Mexican president Miguel de la Madrid proposed an ambitious undertaking: the creation of a comprehensive rural *cadastre* (property register) complete with maps of land plots, each at a scale of 1:50,000. The project, in part intended to clarify the boundaries and holdings of Mexico's many *ejidos* (inalienable concessions of land granted to communities by the state), proved to be more ambitious than he imagined and was quickly abandoned. Mexico's dire economic circumstances played some role in the undertaking's demise but so too did the complicated agrarian reality that, in part, had been the impetus for the project in the first place. Like previous administrations, federal officials found a significant disparity between what appeared in surviving land grant records and what existed on the ground. Over the course of the previous seventy years, ejido lands had been illegally sold, rented, divided, and occupied; different petitioners had been granted the same lands; ejidatarios had migrated away and others had assumed possession of their lands; and lands assigned to a community in one municipality were located on land under the jurisdiction of another.[1] An array of everyday acts, a relentless underground economy, had foiled his adminis-

---

1 Cambrezy and Marchal, *Crónicas de un territorio fraccionado*, 133–34, 157. See also Nuijten, *Power, Community, and the State*.

tration's efforts to carry out one of the most fundamental tasks of the modern state: to account for and regulate landed property and assume control over the space of the state.

The failure of de la Madrid's efforts are all the more remarkable given that the manner in which the granting of ejido land took place after Mexico's revolution had itself been, in part, an attempt to overcome such proprietarial opacity. Those postrevolutionary administrations, in turn, had sought to overcome the kind of spatial ambiguity that had plagued *their* predecessors: the nineteenth-century liberal state-builders who attempted to divide communal lands, institute a regime of simple fee-hold property, attract foreign investment, and resolve land conflicts between villages, municipalities, and states. Although de la Madrid may not have known it, when he ordered the creation of a comprehensive cadastre he was reenacting a drama all too common in Mexican history: a spatial drama of state fixations and fugitive landscapes.

I. STAGE SPACES

The history of the modern Mexican state is inextricably entwined with the space it has not only occupied but actively produced. This study examines one particular dimension of that history and relationship: the contested, dialectical, and social (not merely technical) processes by which explorers, surveyors, and cartographers attempted to define, codify, and naturalize space in cooperation and struggle with the people they encountered in the field. In their *longue durée* analysis of English state formation, Philip Corrigan and Derek Sayer argued that the creation of the modern abstraction of the political state is a process of normalizing and naturalizing "what are in fact ontological and epistemological premises of a particular and historical form of social order."[2] In other words, they suggested, the process of state formation is in part one of "defining, mapping, [and] naming 'reality.'"[3]

In what follows, I take this final observation literally, by analyzing the cartographic routines through which both the Mexican state and the space it occupied were produced and rendered natural. Spatial and car-

2 Corrigan and Sayer, *The Great Arch*, 4.
3 Ibid., 141–42.

tographic metaphors ("mapping" and "space" are conspicuous examples) have gained widespread prominence in an array of academic disciplines. Their application is laudable in as much as it *may* reflect an increased sensitivity to space in critical theory. However, there is a very real danger that a proliferating and metaphorical promiscuity may give such words little more than a trendy banality, sapping them of any critical meaning. Even worse, such uncritical usage may inadvertently imply that both space and cartography are themselves transparent and neutral.[4] There is more than a little irony here, given the concerted efforts made by scholars in recent years to overturn such positivistic notions of space and maps. As they have sharply observed, a consuming concern with process and progress privileges *time* (and its institutional manifestation, history) as the dimension of critical engagement rather than *space* (and its institutional manifestation, geography).[5] Space tends to be perceived as a static and neutral category, a prepolitical object, and little more than a passive stage upon which historical subjects play assigned roles. "Our gaze," Paul Carter writes, "sees *through the space of history*, as if it was never there."[6] Such a lack of perspective is problematic.

In the first place, stage spaces deny certain kinds of agency: the places people have actively created—the transformations of *space* into *place*—appear preformed and preordained, detached from meaning and experience.[7] In reality of course, people make their own geography as well as

4 For cautionary remarks, see Smith and Katz, "Grounding Metaphor," and Mitchell, "Different Diasporas and the Hype of Hybridity." See also Turnbull, *Maps Are Territories*. The conscious and strategic, if still problematic, use of spatial metaphors is typical of much of the writing of Louis Althusser and Michel Foucault. See Althusser and Balibar, *Reading Capital*, and Foucault, "Questions of Geography."

5 See Harvey, *The Condition of Postmodernity*, esp. part 3; Massey, "Politics and Space/Time"; Ross, *The Emergence of Social Space*; and Soja, *Postmodern Geographies*. Soja's oft-cited call for the reassertion of space in critical theory is meant to suggest that space has been left out of the equation by being too readily accepted as unproblematically *already there*. There is a venerable radical tradition in geography upon which these authors build, running from Reclus's *L'Homme et la terre* and Kropotkin's "What Geography Ought to Be" to Lefebvre's *The Production of Space* and Harvey's *Social Justice and the City*. Structuralist studies of spatial production are complemented by the humanist tradition in geography which in its own way denaturalized space. See the classic work of Tuan, *Space and Place*.

6 Carter, *The Road to Botany Bay*, xiv. My emphasis.

7 The distinction between "space" and "place" is a common one, and the liter-

their own history.[8] On the flat field of the coordinated grid, they do neither: place is timeless; history is staged. Agency, when it appears at all, returns as an apologetic for the exercise of power, such as in the settler discourses so trenchantly unsettled by Frantz Fanon: "The settler makes history; his life is an epoch, an Odyssey. He is the absolute beginning:

---

ature is vast. While a variety of definitions exist, a basic one would be that "place" is space to which meaning has been ascribed and endowed with value. For this definition see Tuan, *Space and Place,* and Carter, Donald, and Squires, eds., *Space and Place.* Space and place have served as foundational points of departure for much of the new literature in cultural geography. For overviews, see Gregory, *Geographical Imaginations,* and Mitchell, *Cultural Geography.* Anthropologists' careful thinking about place making can be found in Gupta and Ferguson, eds., *Culture, Power, Place;* Wade, *Blackness and Race Mixture;* Feld and Basso, eds., *Senses of Place;* and Basso, *Wisdom Sits in Places.* Excellent place-sensitive historical studies that I have found useful include Dening, *Islands and Beaches;* Carter, *The Road to Botany Bay;* Faragher, *Sugar Creek;* Richardson, *Possessions;* Roldán, *Blood and Fire;* and Appelbaum, *Muddied Waters.* The prominence of space and place as organizing units of analysis in recent years is not surprising given the transformations wrought on social, political, cultural, and economic life by globalization, the Internet, and flexible accumulation. Concerns with the uneven development of capitalism and a perceived spatial and cultural homogenization (even if globalization rarely flattens differences between places to the degree some lament) has generated a dynamic literature on the relationship between the global and the local. For particularly astute and careful discussions of the relationship between space and place and the (often uncritical) use and application of the terms in battles over globalization, see Massey, "Places and Their Pasts," and Massey, *Space, Place and Gender.* Compare her analysis with Harvey, "Between Space and Time," who sees the fetishization of place and the local as complicit with late capitalism. Massey suggests quite convincingly that place-*based* rather than place-*bound* social movements need not fall prey to the reactionary, exclusionary forces that Harvey fears. For a sweeping philosophical survey of space and place, see Casey, *The Fate of Place.*

8 A comprehensive bibliography is not possible here, but the point is emphasized with particular force in two works straddling the twentieth century: Reclus, *L'Homme et la terre,* and Said, *Orientalism.* Said consistently located geography at the center of his analyses of colonialist thought and practice. The unifying premise of the array of practices that Said identified as orientalism was geographic: the collapsing of over half the globe into a single unit of analysis, understood as having a coherence and essence. See ibid., esp. 49–72; Said, *Culture and Imperialism.* For a superb social, as well as literary, analysis of the making and workings of vernacular geographies, see Pred, *Lost Words and Lost Worlds.* Obviously the work of Fernand Braudel is seminal, yet for all his sensitivity to geography Braudel suggested repeatedly that human history was in the last instance *determined* by geography. See esp. Braudel, *La Méditerranée.*

'This land was created by us'; he is the unceasing cause: 'If we leave, all is lost, and the country will go back to the Middle Ages.'"[9] On the stage space, *only* the settler makes history. In other words, as space becomes a stage, history becomes teleology. The ambiguities of (and struggles in) history are reconciled and suppressed through spatial order as the open-ended yields to the inevitable. The complexity, contingency, messiness, and irony that *is* human history; the struggles for, and alternative visions of, a better social life; the myriad ways of organizing and conceiving space; the spatial practices and relationships that were transformed in the process of primitive accumulation and state formation; and, not least of all, the techniques and technologies of domination—all are flattened and neutralized in the teleological quest for legitimacy, foundational coherence, and the naturalization of the social world.

Thus, an overweening emphasis on history at the expense of space is, ironically enough, ahistorical. Space does not merely display itself to the world, as if it were somehow ontologically prior to the cultural and semiotic codes through which its existence is expressed. Such myths of mimesis turn the historical into the natural, concealing its social, cultural, and political underpinnings. "Space," remarked Henri Lefebvre, "is produced [and] if there is a productive process, then we are dealing with *history*."[10] And with power: "[T]o talk in terms of space," Michel Foucault observed, "to trace the forms of implantation, delimitation and demarcation of objects, the modes of tabulation, the organisation of domains meant the throwing into relief of processes—historical ones, needless to say—of power."[11] Yet by freezing the spatial axis this historical process of production and its link to power is rendered invisible.

The very understanding of space as a stage *has* a history, one inextricably linked to the social abstraction of commodity exchange and the political abstraction of the modern, territorial state.[12] A fundamental stage in capitalist development was the development of the idea of the

9 Fanon, *The Wretched of the Earth*, 51. For more recent attempts to skewer such legitimation histories, see Carter, *The Road to Botany Bay*, and Trouillot, *Silencing the Past*, esp. chap. 4.

10 Lefebvre, *The Production of Space*, 46. Emphasis in the original.

11 Foucault, "Questions of Geography," 70.

12 The following paragraph draws upon Lefebvre, *The Production of Space*; Cosgrove, "Prospect, Perspective and the Landscape Idea"; Harvey, *The Condition of Postmodernity*, esp. part 3; Edgerton, *The Renaissance Rediscovery of Linear Perspective*; and Jay, *Downcast Eyes*.

stage itself. The sixteenth-century application of Euclidean principles of geometry to spatial representation in order to create a "realist illusion" of three-dimensional space on a two-dimensional surface gave artistic expression to a developing new "way of seeing." The geometry that structured perspectival space was itself critical to the growth of activities intimately linked to modern capitalism such as double-entry book-keeping, land surveying, and the production of real property.[13] Just as crucial, because perspectival space was founded upon geometric principles, it was assumed to merely reflect inherent properties of space itself such that a new way of seeing, inextricable from political, economic, and social transformations of the time, became *the* way of seeing.[14] Space— now subject to the universal laws of science, statecraft, and political economy—acquired a scientific and factual existence as an observable object detached from meaning, experience, and politics. Situated within a web of uniform and mathematically configured coordinates, space became self-evident, a socially and historically flat surface amenable to circulation, possession, and control.[15] All the world was now a stage, as William Shakespeare proclaimed while his players performed at the aptly named Globe Theatre.

The connection between the stage space and cartography is an intimate one. The orderly system of linear coordinates is modern cartography's graticule—the epistemological and methodological geometric grid of longitude and latitude imagined to envelope the globe. Once coordinated, all space became *already there*, its reality predicted by the global coordinates that posited its very existence. Modern cartography, founded upon the same geometric and mathematical principles as perspectival space, took form as a supposedly objective science mediating

13 Cosgrove, "Prospect, Perspective and the Landscape Idea"; Harvey, *The Condition of Postmodernity*, 245. See also Rotman, "The Technology of Mathematical Persuasion."

14 Cosgrove, "Prospect, Perspective and the Landscape Idea," 51. More broadly on these transformations, although still with a sense of the spatial revolution they both required and facilitated, see Anderson, *Lineages of the Absolutist State*.

15 Paul Carter has in fact argued that the "very idea of invasion and colonization presupposed a theatrical conception of space." See Carter, *The Lie of the Land*, 365. See also O'Gorman, *The Invention of America*; Hillis, "The Power of the Disembodied Imagination"; Edgerton, *The Renaissance Rediscovery of Linear Perspective*; and Agnew, *Worlds Apart*.

between spatial reality and human perception of that reality. Its products—maps—acquired a disembodied purity, functioning as transparent windows onto a preexisting space. Yet, as Nietzsche sarcastically reminded his readers, there is no "immaculate perception."[16] Maps are no more transparent than language which, *pace* the literary realists, carries with it an array of normative assumptions and ideological premises.[17] The power of the myth of mimesis is its capacity to obscure such assumptions and premises—such interests—behind a facade of both objectivity and neutrality.[18] But if a map reflects anything, it is the relationship between modes of representation and the material practices of power.

II. A SPATIAL HISTORY OF MEXICO

While students of modern Mexican history have devoted relatively little attention to mapmaking, surveying, and exploration, the same cannot be said of the various administrations that have ruled (or aspired to rule) Mexico since its independence in 1821.[19] State officials, bureau-

16 Quoted from Jay, *Downcast Eyes*, 191.

17 For critiques of literary realism, see Eagleton, *Literary Theory*, and White, *The Content of the Form.*

18 This has been most persistently articulated in the work of the late J. B. Harley. See his collected essays in Harley, *The New Nature of Maps*, and his "Rereading the Maps of the Columbian Encounter." For a useful critique of Harley and his use of Foucault and Derrida, see Belyea, "Images of Power." For excellent works that historically ground, and complicate, Harley's theoretical articulations, see Mundy, *The Mapping of New Spain*; Edney, *Mapping an Empire*; Burnett, *Masters of All They Surveyed*; and Michael, "Separating the Yam from the Boulder."

19 Exceptions include Holden, *Mexico and the Survey of Public Lands*; García Martínez, "La Comisión Geográfico-Exploradora"; Rebert, *La Gran Línea*; Mendoza Vargas, ed., *Mexico a través de los mapas*; Mendoza Vargas, "Historia de la geografía en México"; and Tutino, "Agrarian Social Change and Peasant Rebellion." The colonial period has attracted more attention. For a sampling, see Acuña, ed., *Relaciones geográficas del siglo XVI*; León-Portillo and Aguilera, *Mapa de México-Tenochtitlán*; Mundy, *The Mapping of New Spain*; Gruzinski, *The Conquest of Mexico*; Trabulse, *Cartografía mexicana*; Mignolo, "Colonial Situations, Geographical Discourses, and Territorial Representations"; Mignolo, *The Darker Side of the Renaissance*, esp. part 3; and Aguilar Robledo, "Land Use, Land Tenure, and Environmental Change." For an overview of this work, see Craib, "Cartography and Power in the Conquest and Creation of New Spain."

crats, and military personnel in independent Mexico relied heavily upon sets of cartographic routines—exploration, surveying, place-naming, and mapmaking—in order to rule more effectively. More than mere instruments of statecraft, such cartographic routines constitute a significant point of reference for understanding an entire modality and methodology of rule. The activities of, say, the Comisión Geográfico-Exploradora (CGE), Mexico's first federal mapping and exploring agency and the subject of chapters 4 and 5, reveal as much about the culture of rule as do policy statements and legislative decrees. Indeed, their consuming preoccupation with spatial order, scientific rigor, and visuality were part of a broader rationale of rule premised upon a principle basic to the art, theory, and practice of modern government: universal fixity.

Cartographic routines held a simple but significant promise: they would give space a stable signification, permitting it to be more effectively appropriated, transformed, and regulated. They would, in the parlance of the time, fix [*fijar*] the land as a stable, visible, and readable stage. On the one hand, cartographic practices produced material texts *about space* in the form of maps, titles, deeds, and descriptions that could be archived and given the force of law. At the same time, they produced space itself *as a text* through the inscription of lines, points, plots, and place-names. In fact, agrarian bureaucrats, development experts, and an array of federal and regional officials in modern Mexico verily obsessed over spatial fixity of this sort. In order to stress this obsession with permanence, I refer to the various cartographic projects they promoted—the privatization of communal lands; the delineation and archiving of village, municipal, state, and national boundaries; the determination of watercourses and riparian rights, to name only a few—as *state fixations*.

Representational fixity, particularly at the national level, had significant symbolic value. The cumulative effect of practices of mapping and surveying gave an otherwise fragmented polity an aesthetic and visual unity and an imagined entity a very material tangibility.[20] The symbolic potential of cartographic representation, I argue in chapter 1, pushed various administrations to persistently pursue the creation of a *carta*

20 On the state as the structural, metaphysical effect of a multitude of disciplinary formations (many of which are spatial), see Mitchell, "The Limits of the State," and Mitchell, *Colonising Egypt*. See also Philip Abram's classic statement in "Notes on the Difficulty of Studying the State."

general (national map) of the Republic in the immediate aftermath of independence. As territorial loss and foreign invasions punctuated the passage of time, national maps acquired even more importance as a means to set the bounds of territorial sovereignty and to provide a textual tangibility to an otherwise metaphysical entity, effectively helping to create that which they purported only to represent. They also functioned to legitimate a newly ascendant elite's claims to rule: "to represent," after all, "conflate[s] politics and poetics" in that it signifies both to speak *of* and to speak *for*.[21] National maps did not simply imagine the nation-state into existence, but they did function as a means through which such an object could be more effectively imagined, propagated, and circulated; circulated, as I suggest in chapter 5, not only to one's purported subjects, but to foreign investors eager to see an image *representative* of the political stability and spatial predictability necessary for profitable investment.

On a more practical level, exploration, mapping, and surveying became the means by which to identify and assume control over resources, to reconfigure property relations, and to generate knowledge of the territory. These activities were instrumental to the process of territorial integration, to the degree that one could plausibly argue that the state and cartography are reciprocally constitutive.[22] In Mexico, cartographic projects multiplied rapidly in the wake of the Mexican-American War and the promulgation of the Constitution of 1857. After decades of *pronunciamientos*, invasions, territorial amputation, and internal wrangling, the liberal state sought to centralize control and attain some modicum of stability. Mundane activities such as exploration and surveying assumed paramount importance: the information they provided would help produce the kind of mapped, official knowledge so essential to the effective rule of disparate regions. Such activities, for example, would enable agencies to locate and manage resources, mediate claims over land and water, and establish control without depending upon local knowledge.

For these reasons alone, the absence in the historiography of modern

21 Comaroff and Comaroff, *Of Revelation and Revolution*, 1:15; see also Agnew, *Worlds Apart*, 102.
22 See Wood, *The Power of Maps;* Escolar, "Exploration, Cartography, and the Modernization of State Power"; and Kain and Baigent, *The Cadastral Map in the Service of the State.*

Mexico of sustained analysis of surveying and mapping is remarkable; even more so if one considers the pivotal role ascribed to such practices as causal factors in Mexico's twentieth-century revolution.[23] The significance of the figure of the surveyor in modern Mexican history cannot, in my mind, be stressed enough. As I argue in chapters 3 and 7, legislative decrees and gubernatorial speeches did not divide lands prior to the revolution, nor did they create ejidos after it. Surveyors did. Surveyors were neither passive extensions of objective instruments nor an homogeneous and transparent group of lackeys in the service of the state or landlords. Be that as it may, they did have one thing in common: they often appeared in rural areas as intermediaries between an abstract state (and its policies) and local populations who were affected by those policies. People experience "the state" as they experience "the market" or "capitalism," not as a broad abstraction but as a series of manifestations with a very human face: judges, notary publics, police squads, tax collectors. And surveyors. In rural Mexico, both before and after the revolution, surveyors were among the most prominent figures through which villagers *experienced* something known as "the state."

At the same time, if villagers often experienced the state in the form of the surveyor, federal officials and a burgeoning bureaucracy "saw" or came to know the countryside through the surveyor. This was particularly the case with the federal military traverse surveyors (examined in chapter 4) who combined surveying with exploration. Much of Mexico, as both topographic space and natural "resource," resided far beyond the horizon of official knowledge. The need to explore these realms appears with mantric regularity in official rhetoric throughout the nineteenth, and well into the twentieth, centuries. Mexico's infamous nineteenth-century dictator, Porfirio Díaz, assigned this task to the military surveyors of the CGE. Charged in part with creating a "perfect map" of the Republic, the CGE surveyors traversed the countryside gathering the data necessary to produce *structured knowledge* of the physical

23 An exception is Holden's *Mexico and the Survey of Public Lands,* which demonstrated what could be gained from such investigations. His work revealed that the standard narrative of land-hungry survey companies uniformly dispossessing villagers, with the support of President Porfirio Díaz, needs to be revised. A recent call for more attention to be paid to the process of land division is Kourí, "Interpreting the Expropriation of Indian Pueblo Lands." A good start is his "The Business of the Land," on land divisions in late-nineteenth-century Papantla, Veracruz.

topography in the form of maps. As I suggest in chapter 4, the dynamic and radically *un*stable activities of CGE surveyors and explorers created a very *stable* image of the topography, reducing a complex world to manageable proportions and quantifiable configurations. Their grounded views became *overviews* that permitted state officials to see at the proper scale and administrate more effectively.[24] However, to stop the analysis there would miss a key, if initially unintended, consequence of their activities: through their explorations and interactions CGE surveyors simultaneously garnered an enormous amount of *situated knowledge* of the social and political, as well as the physical, topography of the countryside. The deeply contextualized understandings they acquired of the specific regions and localities they mapped made them prime candidates for positions of political power in those very areas.

The surveyor is a necessary protagonist in this story but not the only one. A mapped Mexico did not result solely from the labor of military and civilian surveyors or the meticulous work of metropolitan cartographers. Throughout this book I explore the fundamental role that local people (primarily rural cultivators in the state of Veracruz) played in the surveying and mapping of the countryside. Local people were not fleeting mirages on the surveyor's horizon but agents in their own, and Mexico's, spatial history. They did not stand idly by when surveyors came into view; but nor can their roles be reduced to something as stereotypically simplistic as romantic resisters of some cartographic juggernaut. Not only did they consistently engage the surveyors responsible for stabilizing the ground on which they stood, but their agrarian practices, conceptions of history and geography, and local politics all radically complicated and reshaped the projects surveyors had been assigned.

In point of fact, state fixations all too often ended up as state frustrations. *On* the ground, fantasies of fixity *ran* aground. Regional officials, surveyors, and military mapmakers did not encounter (nor did they expect to) the blank spaces so typical of the imperial imagination. They encountered the kinds of places their own work was designed to both

24 The process of reducing a complex world through such state simplifications has been analyzed with great insight by James C. Scott in his *Seeing Like a State*. My study is significantly indebted to Scott's book, although I would note that the modern Mexican state was at no time in its history a "high modernist" one of the kind that garners Scott's attention. See also de Certeau, *The Practice of Everyday Life*, esp. part 3, and Burnett, *Masters of All They Surveyed*.

reconcile and supersede. They confronted what I will call *fugitive land-scapes.*[25] In their traverse and property surveys, they frequently found themselves in lands characterized by multiple political jurisdictions and use rights, indeterminate borders and inconsistent place-names, and highly contextualized systems of tenure and property. These were intensely local settings—not land*scapes* at all, in fact, but places created and recreated through the prisms of memory, practical wisdom, use, and collective decision making rather than the lens of instrumen-tation.[26]

Yet, to suggest that such landscapes foiled the dream of universal fix-ity sought by the liberal state—one which would make property rights, laws and identities in any given place precisely like those in any other and thus, like the market, utterly place*less*—is not to argue that villagers had little or no sense of fixity in their own right. Villagers in central Veracruz, for example, were not the prescient antecedents of our con-temporary champions of "third" or "in-between" spaces, reveling in antiessentialist counternarratives. They proffered their own fixations regarding property, territory, identity, and history. Indeed, this is pre-cisely why those being "mapped" did not simply ignore or acquiesce and why surveys were so often very charged encounters. Surveying may be a scientific practice, but it is not carried out in an empty or controlled environment. Surveyors, before and after the revolution, quickly be-came enmeshed in the local histories, conflicts, and contexts that were theoretically to remain extraneous to their work. And they found them-selves to be the objects of intense scrutiny, constantly subject to the in-fluences, pressures, and threats of an array of rural inhabitants. Large landowners, for example, feared formal titling procedures and sought to limit the power of a surveyor's report. Villagers, meanwhile, were rarely mystified by the first person they encountered carrying arcane instru-ments. They knew all too well that objective instruments wrought po-litical and social consequences. More to the point, as I show in chapter 3, in many instances village officials *were* the surveyors, a useful re-minder that "the state" and "the local" were hardly mutually exclusive

25 On "fugitivity," see Fritszche, *Reading Berlin, 1900;* Berger, *Ways of Seeing;* and Deleuze and Guattari, *A Thousand Plateaus,* esp. chap. 12.

26 On the powerful asymmetries involved in the term "landscape," see Wil-liams, *The Country and the City.* I thank Aldo Lauria-Santiago for his comments on these issues.

domains. But even when the surveyors came from without, villagers did not necessarily either comply with or resist the various projects these individuals had been charged with completing. At times, of course, they did, but in many instances they appropriated and rearticulated various aspects of these projects for their own purposes. At the same time, surveyors were dependent upon the very vernacular images and conceptions their projects were designed to replace, and their work was frequently characterized by the very traits they attributed to the villagers: localism, petty politics, and self-interest.[27] The spatial creation of Mexico was a much more ambivalent and dialectical process than one of some state juggernaut imposing its visions upon an either quiescent or intransigent countryside.[28]

In sum, it was through an array of everyday activities and processes, through struggles and accommodations played out on multiple levels and registers, that a spatial history of the state *took place*. Activities such as property surveys, the explorations of the CGE, or the boundary delineations conducted by village officials may lack the narrative force of the rebellions and revolutions that attract so much scholarly attention; they may have been less *extraordinary*, but they were no less dramatic. Part of "an epic of the ordinary," they were fraught with far-reaching consequences.[29] It was, after all, through these bureaucratic encounters, quotidian interactions, and documentary exchanges that spaces were (re)assigned meanings and names, ordered and divided, naturalized and signified, and, at least theoretically, constituted and regulated. Little wonder that as conflicts over land and political autonomy progressed in stride with liberal state formation, surveyors rather than soldiers appeared as the most troubling (or promising) figures on the horizon; measuring chains, as well as rifles, became talismans of power; and titles and maps became potent weapons.

27 For a powerful critique of science's claims to objectivity and rationality, to being somehow less contextual and contingent than the local knowledges it claims to supersede, see Feyerabend, *Against Method*. See also Latour, *Science in Action*, and Lenoir, ed., *Inscribing Science*.

28 The vibrant literature on Mexican state formation has been an obvious source of inspiration. See esp. Joseph and Nugent, eds., *Everyday Forms of State Formation*; Escalante Gonzalbo, *Ciudadanos imaginarios*; Nugent, *Spent Cartridges of Revolution*; Mallon, *Peasant and Nation*; Guardino, *Peasants, Politics and the Formation of Mexico's National State*; and Vaughan, *Cultural Politics in Revolution*.

29 Comaroff and Comaroff, *Of Revelation and Revolution*, 2:29.

III. METHOD AND STRUCTURE

This is not a history of maps, nor of landscapes. It is a history of human beings, and I have sought throughout this study to ground my analysis in stories from national, regional, and local archives that seem illustrative of the larger trends that interest me. The focal point is the state of Veracruz, a state whose strategic, agricultural, and industrial importance in the nineteenth and early twentieth centuries gave it a particular geographical preeminence in the minds of national administrations. Not surprisingly, it also figured prominently in the wide array of cartographic projects that punctuated, lined, and named the Mexican landscape.

What follows is not a singular, unified narrative. It is, rather, a series of relatively independent yet related and chronologically overlapping essays that address different federal and regional projects concerned with spatial fixity and thus contingent upon exploration, mapping, and surveying. I chose such a strategy in part because the amount and diversity of cartographic material produced in nineteenth- and early-twentieth-century Mexico preclude any exhaustive examination.[30] At the same time, I wanted to move beyond an analysis of any one project in order to capture the sheer breadth, heterogeneity, and pervasiveness of cartographically based, state-promoted projects of the period. Finally, I wanted to emphasize that a mapped Mexico resulted from a variety of loosely related and relatively diffuse projects rather than from some coherent, unified plan. This is not to suggest that such dispersed and fragmentary processes did not have powerful and systematic effects.[31] While not all of a piece, cartographic projects did nonetheless have a cumulative effect, one that propagated the idea, increased the presence,

30 I do not address, for example, the surveying and mapping of Mexico's public lands. An excellent study on the subject already exists in the form of Holden's *Mexico and the Survey of Public Lands*. Nor do I look at the surveying and mapping of Mexico's national borders, an intentional omission on my part because national borders have been overly fetishized in much of the literature on surveying and mapping. Rebert, *La Gran Línea;* Hall, ed., *Drawing the Borderline;* and de Vos, *Las fronteras de la frontera sur,* offer useful studies of Mexican national boundary formation.

31 This is a basic premise of Michel Foucault's notion of governmentality. See Burchell, Gordon, and Miller, eds., *The Foucault Effect.* See also Ferguson, *The Anti-Politics Machine.*

and buttressed the weight of the state: on the ground, by providing officials with both structured and situated knowledge of the regions they were assigned to administer, and on paper through the production of maps and hefty tomes of geographic and statistical data.

Rather, the point is that *product* should not be mistaken for *process*. The unitary and smooth facade of any image can serve to sever it from the multitude of conflicts, confrontations, and contingencies that went into its very construction. As a consequence, the final image—and, importantly, *a self-image*—appears to simply confirm the inevitability of a given historical moment and spatial expression. To see some master plan (some invisible hand) behind it all is to confuse intention with effect and serves to remove the final image from the quotidian historical and social (not merely technical) processes which conditioned its creation. More to the point, it is to privilege and perpetuate a reified and teleological *stage* space, and in this case, *state* space, one that was always in a process of becoming.

In chapter 1 I examine issues of representation and national sovereignty in nineteenth-century Mexico through the perspective of the first published national map of independent Mexico, Antonio García Cubas's 1858 *Carta general de la República Mexicana*. I suggest that national cartographic projects such as García Cubas's—and the ways in which they connected history and geography—played a critical role in the search for legitimacy and order in the wake of the Mexican-American War.

The next two chapters shift the emphasis from national to individual sovereignty and move from the stuffy halls of the Sociedad Mexicana de Geografía y Estadística to the temperate slopes of the *sierra veracruzana* in eastern Mexico. I turn my attention to the state's attempts to survey and divide communal lands in the latter half of the nineteenth century. Before lands could be divided, village and municipal borders had to be determined. In chapter 2 I focus on those attempts to delineate and map boundaries, emphasizing two issues that made this task extremely difficult for officials: a lack of practical and textual knowledge of the ground and the ways in which new attempts at border fixity conflicted with agrarian practice and custom. In chapter 3 I move to the process of land division itself. A close examination of interactions among surveyors, villagers, and landlords suggests that resistance to the land divisions, when it existed, was not the result of an innate antiliberalism on the part of villagers but rather of their experiences with practical issues of implementation.

Chapters 4 and 5 oscillate between national and regional levels of analysis by examining the works of the Comisión Geográfico-Exploradora (1877–1914), based for most of its existence in Xalapa, Veracruz. Chapter 4 looks at the institution's first two decades of life. I emphasize the dominant role the federal military played in the agency and draw attention to the critical importance that traverse surveying and exploration, and the resulting "structured" and "situated" knowledge they produced, played in the consolidation of Porfirian rule and statecraft. Chapter 5 follows the agency through the last decades of its existence, focusing on the relationship between spatial order and ideas and representations of Porfirian progress.

In chapter 6 I examine a single conflict over water rights between a foreign mine owner and a village council in a remote canyon of Veracruz. I analyze how both parties used conflicting names on state maps, including those of the CGE, to lay claim to a contested waterway. By analyzing the way in which federal officials mediated the dispute, and the ways in which the disputants themselves relied upon various images and names to make their claims, I show *why* the fixing of place-names proved basic to the centralization of state power. But, more important, I demonstrate how the very process of fixing place-names and waterways further legitimated the power of federal bureaucracies whose very legislation had initially spurred such disputes.

Chapter 7 moves the discussion into the postrevolutionary era to trace the continuities and disjunctures in state projects and agrarian relations. I analyze the process of agrarian reform as it developed from the issuance of various decrees during the revolution to the consolidation of the state under Lázaro Cárdenas in the 1930s. The ejido has rightly come to be understood as one of the primary means by which the postrevolutionary state achieved stability and attached the peasantry to it in a vertical relation of reciprocity. But consequences are not the same as intentions. The ejido, and the postrevolutionary state itself, were entities shaped gradually through multiple levels of conflict over both history and geography.

I conclude the book with a short epilogue in which I suggest that the broader arguments contained in this study have some relevance for contemporary issues in Mexico. I return to Miguel de la Madrid's cadastral fantasy and connect it to larger changes shaping the Mexican state, including former president Carlos Salinas de Gortari's recent reforms to the ejido system, which appear as yet another effort in the time-honored quest for the spatial transparency, predictability, and fixity necessary for profit and rule. In an age of "globalization" and neoliberal "integration"—both eminently spatial metaphors—the politics of space, place, and history are as charged as ever.

CHAPTER 1

The Terrain of Tradition

In 1847 the Mexican Liberal politician Mariano Otero, attempting to account for the ease with which "ten or twelve thousand men . . . penetrated from Veracruz to the very capital of the republic," offered a stinging explanation: Mexico did not constitute, nor could it call itself, a nation.[1] Locating the absence of nationhood in the persisting legacies of colonial rule, Otero questioned the degree to which Mexico had moved from colony to modern nation. Such an assertion must have proved disturbing to many, coming as it did a quarter-century after the proclamation of independence from Spanish rule. Certainly the Mexican elite that inherited the mantle of independence in 1821 imagined themselves to be members of a distinctly Mexican nation and state.[2] Yet acts of imagination were not, in and of themselves, powerful enough to sustain Mexico, regardless of how hard "its" leaders imagined, as the turbulent years leading up to and including the Mexican-American War had amply demonstrated. In the wake of the war, the questions that had confronted the republic in 1821 persisted: How would an extensive and complex landscape—and the people inhabiting it—cohere as an intelligible, material unit? How would a new political territory be seen as externally and internally legitimate? And how to demonstrate that a nation, a state, a government were something more than mere conjecture? These were—to borrow a term from philosophy—metaphysical ques-

[1] Otero, "Considerations Relating to the Political and Social Situation of the Mexican Republic in the Year 1847," quotation from Robinson, ed. and trans., *The View from Chapultepec*, 5–31.

[2] The most comprehensive statement to this effect can be found in Brading, *The First America*.

tions, and the methods devised to answer them were part of what could be called a metaphysics of nationalism.

In this chapter I suggest that the disciplines of history and geography were harnessed to answer precisely such questions. In particular I focus on how these two disciplines came together in national mapping projects after the Mexican-American War of 1848. To demonstrate that Mexico was something more than a concept, to legitimate Mexico's spatial and temporal existence, and to make visual arguments about its historical and geographical coherence, intellectuals from the federally backed Sociedad Mexicana de Geografía y Estadística [SMGE] devoted their attentions to the construction of national maps (*cartas generales*) of the Republic. On the purportedly objective surfaces of national maps, they blended history and geography to connect a conceptual space to a narrated place, endowing Mexico with both a textual tangibility and a palpable past. Mexico thus *materialized* on the cartographer's table, a plotted surface upon which the nation-state's past and future could simultaneously unfold.

This chapter is divided into four sections. In part 1 I consider why Mexican officials pursued the construction of a national map. In part 2 I use Antonio García Cubas's 1858 *carta general* of Mexico to show how cartographic science visually naturalized the nation-state. In part 3 I examine how the artistic images that appeared on that same map served to connect the plotted territory to an ideologically saturated portrait of a supposedly quintessentially Mexican landscape. Last, in part 4, I look at how the arbitrary changing of place-names by municipal authorities complicated metropolitan elites' desires to spatially (and cartographically) ground a foundational narrative.

I. VISION

"[A]ll nations have begun as we have, on the road of science," averred Manuel Orozco y Berra in his *Apuntes para la historia de la geografía en México* (1881).[3] That such a statement—revealing as it does the very constructedness of the nation-state—would come from one of Mexico's preeminent geographers is not surprising. Geography proved a key sci-

3 Orozco y Berra, *Apuntes*, 425. All translations are mine unless otherwise noted.

ence in the formation of nineteenth-century nation-states and had a close association with the technical, regulatory needs of those in power. Rising military and economic nationalism spurred the professionalization of geography, its incorporation as a discipline in the halls of higher learning, and the founding of national geographic societies.[4] Latin America's first geographic society—Mexico's Instituto Nacional de Geografía y Estadística (later to become the Sociedad Mexicana de Geografía y Estadística)—had been created in 1833 by then president Valentín Gómez Farías, guided by the belief that the accumulation and production of geographic and statistical knowledge were critical for national development.[5] As the minister of domestic and foreign relations, José María Gutiérrez Estrada, put it two years later, the sciences of geography and statistics were of "extreme importance to the prosperity and good governance of the Nation."[6]

Such concerns ensured that, once established, the Instituto could count on the assistance of the federal government despite the constant shifts in political power that characterized the initial decades of the Mexican Republic. Both Conservatives and Liberals financially supported the institution and the only changes it experienced were in its name.[7] For example, in 1839 the organization was renamed the Comisión de Estadística Militar, at the request of Juan Nepomuceno Almonte, a ranking military officer and son of José María Morelos y Pavón. Now under the auspices of the Ministry of War, the Comisión's employees received a federal salary and were obligated to complete the tasks they were assigned. The shift reflected a sense of urgency in the upper echelons of the Mexican military regarding the collection of geographic

4 See the essays collected in Godlewska and Smith, eds., *Geography and Empire;* Capel, "Institutionalization of Geography and Strategies of Change"; and Harley, "The Map and the Development of the History of Cartography." More broadly, see Said, *Culture and Imperialism.*

5 See Mendoza Vargas, "Historia de la geografía en México," 42–50. The best studies on the history of the Sociedad are two unpublished theses: Mendoza Vargas, "Historia de la geografía en México," and Lozano Meza, "La Sociedad Mexicana de Geografía y Estadística." See also Aznar Barbachano, "Importancia del estudio de la geografía y estadística," and Olavarría y Ferrari, *La Sociedad Mexicana de Geografía y Estadística.*

6 Quotation from Mayer Celis, *Entre el infierno de una realidad,* 86.

7 See Mendoza Vargas, "Historia de la geografía," part 2.

and statistical data as Texas moved increasingly toward secession.[8] A decade later, the organization adopted the name it still has today: the Sociedad Mexicana de Geografía y Estadística.

Statistics and geography were sciences of statecraft. Strongly influenced by the utilitarianism of Bentham and the political economy of Smith, Mexican intellectuals and officials saw statistical inquiry as a means not only to control the vicissitudes of reality but to shape it.[9] With the capacity to measure and compare came the capacity to plan, modify, and transform economies, spaces, and populations. The urge to measure and plan also structured the understanding of geography at the time but, at least at mid-century, geography meant something very specific to members of the SMGE and the federal government: the creation of a national map of the Republic.[10] Why the emphasis upon a national map? Certainly there were very practical concerns related to governance, particularly in the early years of the Republic. Without a reliable national map the new government could hardly begin to conceive of, let alone carry out, any political reorganization of the territory. This would prove a constant source of concern in the recurring territorial reconstructions of the country's politicoterritorial divisions by federalists and centrists, each of whom had their own politicoadministrative geographies.[11] A national map could also prove useful in the war against fiscal chaos, administrative fragmentation, and regional politics in that a variety of local and regional statistical information, and maps could be compiled and incorporated into a master map.[12] More important, perhaps, a

8 Mayer Celis, *Entre el infierno de una realidad*, 91.

9 For detailed studies on statistics and the Mexican state in the first half of the nineteenth century, see Mayer Celis, *Entre el infierno de una realidad*, and Lozana Meza, "La Sociedad Mexicana de Geografía y Estadística." For a good example of nineteenth-century statistical concerns, see García Cubas, "Informe sobre el estado actual de la Estadística nacional."

10 On the SMGE's obsession with a national map, see Lozana Meza, "La Sociedad Mexicana de Geografía y Estadística," 125–26.

11 See O'Gorman, *Historia de las divisiones territoriales*; McGowan, *Geografía político administrativa de la Reforma*; San Juan Victoria and Velázquez Ramírez, "La formación del estado y las políticas económicas (1821–1880)"; and Jiménez Mora, "El proyecto de división territorial de Manuel Orozco y Berra."

12 On the abundance and precision of state maps, see Moral, "Condiciones del trabajo geográfico." See also Orozco y Berra, *Apuntes*, and Anna, *Forging Mexico*, 100.

national map of geographic and topographic accuracy could improve the fledgling state's military capacity during a time of both international and domestic uncertainty, at least for the macro-coordination required for national defense. Thus in the 1820s the government created a new course of study in geographic engineering, commissioned individuals to "travel throughout the entire territory and assemble statistics and a geographic map," and composed a national map from the remnants of the Spanish navy's collection of images created for the defense of New Spain.[13]

Such concerns provide an initial explanation for the persistent pursuit of a carta general but not a complete one, particularly as the years progressed. The fact remains that national maps are of such small scale that they often have minimal instrumental value. A military expedition sent to crush a rural rebellion or ward off a foreign invasion across the *mesa central* would find only so much of value in a map of the entire republic. The plotting of routes and planning of tactics required the large-scale topographic maps produced by military engineers on careful traverse surveys through the countryside, not the small-scale political and geographic overviews of a carta general constructed from a compilation of sources. Similarly, development efforts, such as the building of roads that would tie regional economies and politics to a central apparatus, required primarily regional and local maps of various kinds.

Yet until the 1860s the federally subsidized SMGE, as well as its earlier incarnations, still devoted the vast majority of its energies to creating a carta general, one which would be a "faithful expression of the land it represents."[14] Why? A national map had as much iconographic as it did instrumental power. For one, a national map served the very basic function of defining a bounded space within which a newly emergent postimperial elite could purport to assert their power, confirm their continuing status, and legitimate their rights to rule and, in effect, represent.[15] Moreover, a national map symbolically affirmed the political reality of

13 See Mendoza Vargas, "Las opciones geográficas." The quoted text is from Moral, "Condiciones del trabajo geográfico," 3.

14 Actas de la Comisión de Estadística Militar, October 27, 1839, quotation from Mendoza Vargas, "Historia de la geografía," 54–58.

15 On the way in which politics and poetics dovetail in the active verb "to represent," see Comaroff and Comaroff, *Of Revelation and Revolution*, 1:15, and Agnew, *Worlds Apart*, 102.

an entity whose very existence was at the time increasingly in question—a unified and sovereign Mexican nation-state. Rebellions in northern territories, the secession of Texas and then the Yucatán, and regional conflicts all confounded any comforting thoughts of a unified national space and repeatedly raised the specter of total national disintegration. A national map refuted such troublesome realities by visually affirming what supposedly already existed: after all, if a map were simply a mimetic reflection of an objective reality, then a national map by definition presupposed the existence of the nation itself.[16] The still precarious and open-ended process of forging an independent Mexico appeared as authoritatively over, concluded and confirmed. A scale map of a nation-state, which furthered the ideological mirage of neutrality by applying objective mathematical principles to map construction, thus argued backward from the desired conclusion, serving as a model *for*, rather than *of*, what it purportedly represented.[17]

Even simply delineating where Mexico ended and other nations began could be significant at a time when established boundaries and territorial cohesion were increasingly regarded as integral features of the modern nation-state.[18] Indeed, the powerful sway of territoriality as the basis for modern identity and control ensured that geographic science and its primary medium, the map, occupied a place of preeminence in the nationalist repertoire. This was particularly the case by the 1840s. The strident predations of Mexico's northern neighbor, with its fervent faith in Manifest Destiny, left little room or time for what one author has aptly termed "growing pains."[19] In a manner befitting their continentalist convictions, and further evidence of the power of the geographic imagination at the time, U.S. officials relied upon a kind of cartographic determinism to justify their imperial pretensions.[20] Already

16 For critiques of the powerful and persistent notion that maps simply mirror a spatial reality, see the various works of J. B. Harley, collected in *The New Nature of Maps*.

17 See the sophisticated discussion in Thongchai, *Siam Mapped*, esp. chap. 2. On signs as predictive, see Rama, *The Lettered City*.

18 On the increasing role of territorial coherence in the definition of the modern nation-state, see Hobsbawm, *Nations and Nationalism since 1870*; Maier, "Consigning the Twentieth Century to History"; and Sahlins, *Boundaries*.

19 Archer, "Discord, Disjunction, and Reveries of Past and Future Glories," 192.

20 On cartographic determinism and U.S. imperialism, see Craib and Bur-

in 1823 John Quincy Adams had equated geographical proximity with historical providence when he promulgated his so-called ripe apple policy, which argued that Cuba and Puerto Rico were "natural appendages to the North American continent," fated to fall under U.S. control once the proper conditions prevailed.[21] In 1825 U.S. secretary of state Henry Clay took such geographic determinism to an audacious extreme by suggesting to Mexican officials that turning over the northern reaches of Mexico would actually benefit the country by geographically centralizing its capital.[22] By 1844 businessman and Democrat John O'Sullivan could comfortably assert that anyone "who cast a glance over the map of North America" could see that Texas was "a huge fragment, artificially broken off" from the continent to which it naturally belonged.[23] He had little cause for concern: Nature and Nation soon united.

The importance of the carta general took on dramatic significance with the Mexican-American War. While countries such as the United States, England, Spain, and France achieved a degree of self definition through imperial expansion, Mexico's imperative need to construct and present itself as a sovereign, independent nation-state arose in the face of invasion and perceived impotence.[24] García Cubas put it dramatically in his summation of the Armistice of 1847: "[O]ur history is written simply by saying that Mexico and the United States are neighbors. At least France and England are separated by the Channel; between our nation and our neighbor there exists no other border than a simple mathematic line. . . . God help the Republic!"[25] The members of the Comisión de Estadística Militar, in 1848, hinted at the continuing threat months following the armistice when they rhetorically asked, "How can one expect to understand the nation's territorial extension, or consult regarding its defense, without the formation of a general map and one of each State and territory?"[26]

---

nett, "Insular Visions." Succinct overviews of U.S. policy toward Mexico during this period can be found in Zoraida Vázquez and Meyer, *México frente a Estados Unidos*, esp. chaps. 2 and 3; and Schoultz, *Beneath the United States*, chap. 2.

21 Quotation from Sellers, *The Market Revolution*, 100. See also Craib and Burnett, "Insular Visions."

22 Quotation from Meinig, *The Shaping of America*, 2:135.

23 O'Sullivan, "The Texas Question," 424–25.

24 See Riva Palacio, ed., *México a través de los siglos*, introduction to vol. 5.

25 Quotation from Collado, "Antonio García Cubas," 443.

26 Quotation from Mendoza Vargas, "Historia de la geografía," 55.

Under these less than auspicious circumstances the SMGE's new national map, hastily finished in the aftermath of the war and during the initial phases of the boundary demarcation, appeared in 1850.[27] Along with a wealth of statistical information and comparative tables, it included a visual elaboration of the territory lost in the Treaty of Guadalupe Hidalgo, as well as the demarcation of the new international limits between Mexico and the United States. Reflective of the increasing primacy of the visual in the nineteenth century, the image brought an expression of bitterness from General Santa Anna, who for the first time could actually envision the magnitude of territory Mexico had lost.[28] The map never saw publication because of the government's precarious financial condition after the war. Members of the Comisión and the Sociedad sought publishers in the United States and England but found the prices for publication no more accommodating than in Mexico.[29] As a consequence, in 1851 a foreign traveler could still warn others: "[T]here is no complete map of the territory which may be confidently relied upon."[30]

The need for a published and circulated, Mexican-produced, national map became even more pronounced when, in 1854, Mexico lost another portion of its territorial claims as a partial result of a faulty U.S. map. Article 5 of the Treaty of Guadalupe Hidalgo dictated that John Disturnell's 1846 *Mapa de los Estados-Unidos de Méjico* be used to set the boundary line between the two nations.[31] However, perceived defects in the map regarding the location of El Paso and the course of the Río Grande helped justify renewed U.S. territorial claims, culminating in the 1853–1854 Gadsden Purchase.[32] Regardless of the role General Antonio López de Santa Anna and others played in the politics of the Purchase,

27 *Carta General de la República Mexicana formada por la sección de geografía de la Sociedad Mexicana de Geografía y Estadística, con vista de la que arregló la misma sección en el año anterior y demás datos adquiridos posteriormente.*

28 García Cubas, *El libro de mis recuerdos,* 452.

29 Lozano Meza, "La Sociedad Mexicana de Geografía y Estadística," chap. 2.

30 Mayer, *Mexico,* 2:9.

31 "Map of the United Mexican States, as organized and defined by various acts of the Congress of said republic, and constructed according to the best authorities. Revised edition. Published in New York, in 1847, by J. Disturnell." See *Tratado de paz, amistad y límites,* 7–8.

32 Meinig, *The Shaping of America,* 2:152. Issues were further complicated by the fact that different editions of Disturnell's map were appended to the U.S. and

Mexican officials and intellectuals were convinced: Mexico needed an accurate and internationally accepted national map of its own, published and circulated.[33]

But was it enough to merely delineate the nation's territorial extent? Otero, in 1847, observed that it was "useless to point out that the Mexican Republic possesses an immense territory of more than [840,000 square miles]" when Mexico itself lacked a "national spirit."[34] After the war, a new carta general, constructed by García Cubas, would proffer an iconographic image of the state's new parameters and fill that territory with the ghosts of the past, in the process creating an image of a single national spirit.

II. NATURALIZATION

Shortly after the Mexican-American War, Antonio García Cubas (1832–1912) made a name for himself as one of Mexico's leading geographers and cartographers. He began his career in the offices of the Ministry of Colonization and Industry, simultaneously studying engineering at the Colegio Nacional de Minería. Limited by his widowed mother's financial straits, he took longer than usual to finish his degree, eventually graduating in 1865. In the meantime, he worked diligently on various cartographic and geographic projects, spending his free afternoons and evenings in the library of the SMGE and in the private collections of a number of its members.

The corridors of the SMGE, and the pages of its *Boletín*, exposed García Cubas to two generations of intellectuals—both conservative and liberal—who carried on a long tradition of scientific scholarship in Mexico. The Comisión de Estadística Militar had been largely populated by military men of high rank, such as Pedro García Conde, Mariano Arista, Juan Almonte, and Juan Velázquez de León. But by the early 1850s the newly named Sociedad had begun to incorporate an array of well-to-do

the Mexican versions of the treaty. On the mapping of the boundary, see Rebert, *La Gran Línea*, and Tamayo P. de Ham, *La geografía*.

33 On the continuing efforts to publish the Comisión's carta general, see Lozana Meza, "La Sociedad Mexicana de Geografía y Estadística," chap. 2; and Olavarría y Ferrari, *La Sociedad Mexicana de Geografía y Estadística*, 77–80.

34 Quotation from Robinson, ed. and trans., *The View from Chapultepec*, 29–30. See also Otero, *Ensayo sobre el verdadero estado*.

civilian scientists and intellectuals from Mexico City into its ranks, many of them trained at the Colegio Nacional de Minería. Whether military or civilian, the members of the Sociedad constituted a single scientific community: they frequented the same literary and scientific events, bookstores, theaters, and cafes, and lived in fairly close proximity near the Zócalo.[35] Many of them had lived through the War of Independence, and all, obviously, had experienced the humiliating defeat of 1848. As a result of this final experience in particular, and regardless of their political persuasions, they thus shared one more thing in common: an abiding interest in sciences (especially geography and statistics but also increasingly ethnography, linguistics, and history) as utilitarian undertakings crucial to the formation of an integral nation-state.[36]

They thus filled the pages of the Sociedad's *Boletín* with studies of primary importance to both the representation and facilitation of stable governance. As well as contributing an array of statistical treatises, the *Boletín*'s contributors produced studies on the measurement of roads and distances; the standardization of weights and measures; pre-Columbian, colonial, and contemporary history; indigenous languages; regional and local chorographies; and archeological sites and recommendations for their protection. This then was the Sociedad into which García Cubas was inducted as an honorary member in 1856, at the precocious age of twenty-four: a federally funded institution composed of some of Mexico's most prominent military officials and civilian men of letters who conceived of geography, statistics, and history as utilitarian sciences that derived their value from their contribution to Mexican state formation.

Quickly befriended by Orozco y Berra and historian José Fernando Ramírez, García Cubas flourished. Before the end of the decade, he would be widely considered one of Mexico's premier cartographers and geographers, on a par with his elderly mentor, Orozco y Berra. In the coming years, his pictorial-descriptive maps and atlases constituted the most important and well-known images of the Mexican nation-state produced prior to the publication of the maps of the Comisión Geográfico-Exploradora in the last decade of the century. They hung in the halls of power in Mexico City and on the walls of classrooms; they graced the pages of national histories, such as the multivolume *México a*

35 Mayer Celis, *Entre el infierno de una realidad*, chap. 3.

36 See ibid. and Lozano Meza, "La Sociedad Mexicana de Geografía y Estadística," chap. 2.

través de los siglos (1887–1889), and were exported to foreign countries, where they were highly regarded as authoritative sources for publishers of guidebooks.[37] He also wrote numerous "booster" works, designed to promote Mexico abroad as a place for both physical and economic colonization, and a series of geography texts for Mexican schools.[38]

The work that catapulted García Cubas to fame within government circles and the Sociedad was his carta general of 1857. In July 1856 García Cubas showed a number of the members in the Sociedad a national map he had produced based upon his consultation of various maps and atlases.[39] The members of the Sociedad were evidently extremely impressed, and García Cubas published the carta general soon after to wide acclaim.[40] This carta general became the best known national map of Mexico well into the next decade and served as the basis for Orozco y Berra's own cartographic reorganization of the political landscape under the French empire in 1865. What is of particular interest here is the slightly modified version of this carta general that García Cubas created for his 1858 *Atlas geográfico, estadístico e histórico de la República Mexicana* (figure 1).[41]

García Cubas produced the atlas in part to aid the grand projects of an ascendant Liberal regime: colonization, capitalist development, and the disentailment of church and Indian lands. But the atlas, and particularly the modified carta general in its pages, also serves as an exemplary representation of a new nationalist sensibility arising from the

37 See, for example, Conkling, *Appleton's Guide to Mexico*, 101; and Janvier, *The Mexican Guide.*

38 See, for example, García Cubas, *The Republic of México in 1876;* García Cubas, *Mexico;* and García Cubas, *Cuadro geográfico, estadístico, descriptivo, e histórico.* He also made cartographic games for children such as his *Los insurgentes: Juego histórico para niños* (1891), a game about Mexico's War of Independence in which children would follow the routes of the insurgent generals across the face of a map of Mexico. The map for this game is available at the Nettie Lee Benson Latin American Library, University of Texas at Austin.

39 Mendoza Vargas, "Las opciones geográficas," 104–5.

40 *Carta general de la República Mexicana.* On the map's reception, as well as the comments of Orozco y Berra discussed below, see those of minister of development Blas Balcárcel and of Mexico's preeminent scientist, Francisco Díaz Covarrubias, as quoted in Olavarría y Ferrari, *La Sociedad Mexicana de Geografía y Estadística*, 99.

41 "Carta general en major escala," in García Cubas, *Atlas geográfico, estadístico e histórico.*

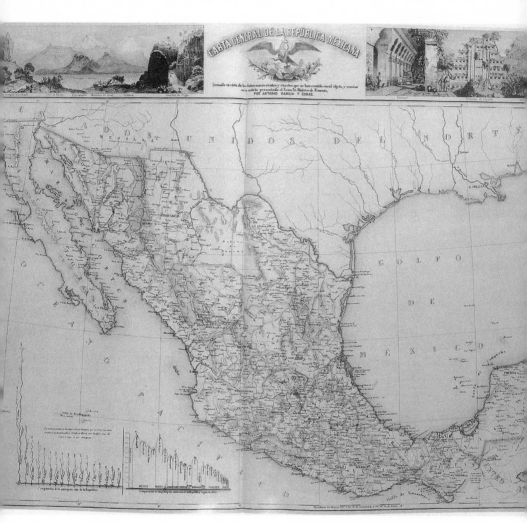

FIGURE I. A metaphysical Mexico: Antonio García Cubas, *Carta general de la República Mexicana*, 1858. Courtesy of the Mapoteca Manuel Orozco y Berra, Mexico City. Photograph by Carmen H. Piña.

Mexican-American War. Here, for the first time, a carta general purported to offer not only a vision of Mexico's geography—of its territorial extent—but of its history. On the surface of the map, history and geography came together to compose Mexico as a coherent historical and geographical entity; that is, as a legitimate nation-state. In one sense the two disciplines came together in García Cubas's own conception of history, which he understood as a geographically descriptive enterprise aimed at discerning how the country literally took shape. His maps and atlases were genealogies of the territory, narrating a kind of property history in which the historical existence of the nation-state was taken as a given and a history of "its" territory was simply recounted. Hence his inclusion in the atlas of a lengthy political genealogy that traced contemporary Mexico's politicohistorical origins back to at least the seventh century and the kingdom of the Toltecs. And thus his devotion of generous space, in his *Atlas geográfico, estadístico e histórico de la República Mexicana*, to the Mexican-American War (which resulted in a massive territorial amputation) and his reduction of the French Intervention (which did not) to a few scanty paragraphs.[42] But history and geography came together in other ways on his carta general, in particular through a careful blend of scientific and artistic images.

To understand how, I begin with García Cubas's cartographic method. What did it mean to be a cartographer in Mexico in the mid-nineteenth century? Any image of a solo explorer, slogging through the brush mopping sweat from his brow, waging warfare against teeming insects while straining under the weight of expensive instruments, would be wide of the mark. García Cubas did very little fieldwork or surveying to construct his maps. Other than his historical map of the battle of Cinco de Mayo, for which he traveled to Puebla and toured the battlefield, all of his maps appear to have been constructed in his office in Mexico City.[43] The city's opposition press picked up on this very point,

42 See the astute analysis in Collado, "Antonio García Cubas." In 1861, the troops of Napoleon III, emperor of France, occupied the port of Veracruz. They marched inland and, after nearly two years of fighting, captured Mexico City. With the support of Mexico's Conservatives, Napoleon III installed Maximilian, the Hapsburg archduke of Austria, as Mexico's monarch. Maximilian ruled until his execution, and the expulsion of the French, by troops of Liberal leader Benito Juárez in 1867.

43 See García Cubas, *El libro de mis recuerdos*, 395; de P. Piña, "Importancia de los trabajos geográficos e históricos."

ridiculing the distinguished cartographer in rhyme: "Without making any stops/ or even moving for a second/ he knows the entire world/ at least by a map."[44] They, for one, understood the artificial conception of reality held by those who learned of the world from the comfort of metropolitan parlors.

García Cubas might have been surprised by their scorn. On one level it was simple: he was not a surveyor. To be a mapmaker in mid-nineteenth century Mexico was to compile images and plot them onto a mathematically ordered surface: a task suited to an office rather than the field. Indeed, García Cubas proudly proclaimed that his maps were based not upon his own fieldwork but upon the "most recent and reliable information," collected from state and municipal governments. The process entailed collecting the maps and then comparing them. It was a rational, rather than empirical, project, dependent upon reason and deduction in lieu of experience and exploration. The presumption was that through careful comparison, regional and national maps could be unified, points determined, and geographical relations corrected. It reflected not only a lack of personnel and money to conduct large triangular surveys, but also a faith in (and fascination with) encyclopedic forms of knowledge construction common at the time. When Orozco y Berra began his multivolume *Diccionario universal de historia y geografía* in 1853, he proudly wrote that his work was one of "compilation and not of creation."[45] The first carta general of 1850 included a statement of authenticity, attesting that in the formation of the map the cartographers had gathered close to three hundred maps of the territory. García Cubas touted his own carta general as being nothing more than the product of careful comparative analysis of "the most exact maps" available at the time.

Regardless of the quantity or quality of maps consulted, map compilation suffered from an irresolvable tension: if the previous maps had been incorrect and inaccurate, what assurances did one have that the map they contributed to building would not suffer similar problems?

44 Quotation from Tenorio-Trillo, *Mexico at the World's Fairs*, 174–75.

45 Quotation from Pérez Rosales, "Manuel Orozco y Berra," 367. These kinds of encyclopedic fetishisms reflected a broader nineteenth-century faith in the presumed unity and totality of knowledge and humankind's capacity to a complete and comprehensive knowledge of physical reality. For a compelling literary perspective, see the analysis in Richards, *The Imperial Archive*.

There were none. A contemporary review by Orozco y Berra of García Cubas's carta general observed that "the work is . . . of simple compilation; it is not perfect and still shows considerable errors."[46] However, he continued, "the indisputable merit of García Cubas's map consists in his reuniting the best existent maps, coordinating them, and bringing them to light, completing for the first time an enterprise that had been impossible for the Sociedad de Geografía, and which in spite of its defects is as of today the only one of its kind and has filled a great space in the geographic science of our nation."[47] As the author here suggests, García Cubas's success came from his compilation of the "best existent maps" into a coordinated, coherent whole. An act of symbolic centralization, it garnered its status not as a result of a series of comprehensive and careful field surveys nor from necessarily correcting previous maps but from the unification of disparate regional maps into a single bound whole. The visual effect minimized variation and rupture, offering instead a projected unity known as the nation-state. The success of the image, and its international legitimacy, derived also from the fact that García Cubas not only reunited a variety of existing maps but also "coordinated" them, as Orozco y Berra's felicitous word choice reveals. He literally coordinated the images by superimposing a graticule—the net of imaginary parallels and meridians thought to envelope the globe and which together provide geographic coordinates—onto his compiled material. Within the graticule, García Cubas positioned Mexico for the first time in relation to the Greenwich meridian rather than the easternmost point on the cathedral in the central plaza of Mexico City, the traditional meridian for Mexican maps.[48] He thus brought Mexico into cartographic consonance with what were then construed to be the icons of advanced civilization, giving it a "modern" spatial sensibility.

But García Cubas's use of the graticule surpassed the mere act of making sure his coordinates were internationally coordinated. While the concept of the graticule has a history, the graticule *itself* is strictly ahistorical in terms of what it delineates: it is, in theory, a reflection of a mathematically derived order, itself supposedly a mirror of the natural order of the universe.[49] In other words, it is something not so much cre-

46 Orozco y Berra, *Apuntes*, 424.
47 Ibid.
48 I thank Héctor Mendoza Vargas for bringing this to my attention.
49 Thus Galileo's assertion that the universe could be understood as a "grand

ated as discovered according to formal rules of mathematical logic. Mexico's location within this timeless matrix made a similar subtle and transhistorical assertion—it was a nation discovered rather than created. Structured by the graticule, the nation-state appeared as an objective reality, existing in advance of its own exploration. Its physical existence predicted by global coordinates, all that remained was to better render its dimensions and internal composition, a process guaranteed by a firm belief in scientific progress.[50] In effect, García Cubas scientifically naturalized the Mexican nation-state through the visual medium of the map.

III. VISUALIZATION

A plotted, scientifically naturalized territory did not make Mexico. While the graticule predicted and structured a given *space*, it did not reveal a *place*.[51] To make Mexico a tangible reality the scientifically derived surface needed to be attached to a visual panorama.[52] Thus, adjoining the graticule, carefully placed so as not to obscure nor blend with the lined surface, lay artistic images that provided a visual, historical, and spatial anchor to the plotted points of the abstract grid.[53] These images visually complemented and amplified the coordinates that covered, and connected, a cartographic Mexico. They gave the scientific image an aesthetic and historical depth, infused a modern methodology with foundational mythology, and reconciled the pervasive nineteenth-century nationalist tension between modernity and authenticity.

To the right of the cartouche, García Cubas reproduced a number of

---

book ... written in the language of mathematics" (as quoted in Rotman, "The Technology of Mathematical Persuasion," 55).

50 For comparative perspectives on the relationship between cartographic graticules and the spatial imagination, see Thongchai, *Siam Mapped;* Carter, *The Lie of the Land;* Burnett, *Masters of All They Surveyed;* and Hillis, "The Power of the Disembodied Imagination."

51 On the distinction between "space" and "place," see the introduction.

52 My thinking on the relationship between plotted points and pictorial representation is indebted to Burnett's careful discussion in *Masters of All They Surveyed,* esp. chap. 3.

53 I have taken the term "spatial anchor" from Basso, *Wisdom Sits in Places,* and adapted it to my own purposes.

FIGURE 2. The primordial landscape: Detail from García Cubas, *Carta general.* Courtesy of the Mapoteca Manuel Orozco y Berra, Mexico City. Photograph by Carmen H. Piña.

popular images of archaeological sites: Palenque, Pirámide de Papantla, Mitla, and Uxmal (figure 2). The images are indicative of an increasing reliance at mid-century upon the pre-Columbian past to improve Mexico's national image. Certainly this was not the *indigemanía* of Porfirian Mexico, when state officials presented Aztec palaces at world's fairs, unveiled statues of Aztec heroes such as Cuauhtémoc on Mexico City's Paseo de la Reforma, and devoted an entire volume of *México a través de los siglos,* the ambitious synthesis of Mexico's past, to the pre-Hispanic era.[54] This kind of neo-Aztecan indigemanía was still inchoate in the 1850s and 1860s. It would be another two decades after the publication of García Cubas's map before lands containing the very archaeological monuments he painted were even exempted from alienation or sale and before an office of Inspección y Conservación de Monumentos Arqueológicos de la República would be founded.[55] Yet neither was this the

54 Riva Palacio, ed., *México a través de los siglos,* esp. vol. 1 by Alfredo Chavero. On these various forms of self-representation, see Tenorio-Trillo, *Mexico at the World's Fairs;* on the statue of Cuauhtémoc see Tenenbaum, "Streetwise History." For the changing political and intellectual context of these conceptions, see Hale, *Liberalism in the Age of Mora,* esp. chap. 7, and Hale, *The Transformation of Liberalism.*

55 On the exemption of lands with monuments, see the circular sent to all state governors in 1877 by the Secretaría de Fomento, Colonización, Industria y Comercio. Reprinted by the government of Veracruz as Circular 25, October 5, 1877, in CLEV—*1876–78,* 146–47. On the office of *Inspección,* see Tenorio-Trillo, *Mexico at the World's Fairs,* 84–89.

Mexico of the 1830s, characterized by a general "indifference to Mexico's indigenous heritage," in which the most conspicuous commentary on Indians in elite writings was their utter absence.[56] By the 1850s articles on pre-Columbian society, history, archaeology, and even cartography began to appear regularly in the SMGE's *Boletín*.[57] An editorialist observed in 1850 that not only did foreign writers provide a distorted view of Mexico by focusing only upon its recent history, they also ignored the cultural achievements of contemporary Mexicans' direct predecessors, who were entitled "to be regarded as the most cultured people the Spaniards found in the New World."[58] Meanwhile, intellectuals from the SMGE drew upon the perspectives of Clavijero and Teresa de Mier to appropriate a generalized indigenous past for historical precedent and priority.

The images in García Cubas's cartouche are part of this moment. While the Nahua would assume increasing prominence on the ancestral pedestal, a number of sedentary indigenous groups that had inhabited the terrain within contemporary Mexican national boundaries—such as the Maya, Tarascans, Zapotecs—were included in the pantheon of Mexican patrimony. In sharp contrast, nonsedentarists such as the Apache and Comanche, from their now predetermined peripheral realms, were deemed treacherous enemies of the state-in-formation. Both Orozco y Berra and García Cubas cast the Indians of the central plateau as sedentarists living the agrarian romance, defending civilization and progress from the incursions of "perfidious, traitorous and cruel" northern tribes.[59] The nonsedentarists were wandering tribes

56 Hale, *Liberalism in the Age of Mora*, 217.

57 By 1860 SMGE members were arguing that archaeological sites should be declared the property of the federal government. See "Dictamen presentado a la Sociedad de Geografía y Estadística por la comisión especial"; and Romero Ramírez, "Proyecto de ley relativo a la conservación de monumentos arqueológicos." See also Orozco y Berra, *Materiales para una cartografía mexicana*.

58 Quoted in Keen, *The Aztec Image in Western Thought*, 412.

59 Orozco y Berra, *Historia antigua y de la conquista de México*; and García Cubas, *Etude géographique, statistique, descriptive, et historique*, quoted in Tenorio-Trillo, *Mexico at the World's Fairs*, 89. On Orozco y Berra's work see the useful discussion in Villoro, *Los grandes momentos del indigenismo en México*, 149–68. Not surprisingly, while Mexican intellectuals labored over Nahua, Tarascan, and Mixtec languages, the study of the Athapaskan languages of the northern frontier was left to foreign intellectuals. See for example the extensive studies of the Atha-

rather than civilizations: with no proper ruins or remnants to take one back in time and no rootedness to satisfy the nostalgia for origins, they were construed as having no history to speak of. A nationalist narration of the passage of time could only begin by envisioning a permanence in space. Connecting contemporary Mexico to a variety of *sedentarist* indigenous pasts portrayed it as a presumably unified territorial entity of historical longevity with a statist tradition, endowing the government's own centralizing tendencies with historical pedigree. Images of disparate archaeological sites, such as the ones in the image here, were thus part of a "selective tradition" created through a mutually reinforcing linkage of historical veracity and geographical priority.[60]

The images also elevated Mexico's cultural capital in an increasingly exoticizing world while drawing attention away from the reality of the contemporary Indian, a problematic issue for both Liberals and Conservatives. Contemporary Indians conjured up images of caste wars and colonial legacies and were understood largely as a "problem" to which Liberal and Conservative alike offered varying solutions: the abolition of communal land tenure, their political incorporation as national citizens, or "assimilation" through European immigration and "whitening." Conservative Francisco Pimentel complained that "in Mexico there is no commonality between whites and Indians; everything is different: physical aspects, language, customs, [and] state of civilization. In Mexico there are two different nations on the same land, and worse still, two nations that are to a point enemies."[61] A Liberal counterpart voiced similar concerns that contemporary Indians were not Mexicans as they "still conserve their own nationality, protected by family and language."[62] The national press could be at times more blunt: according to an editorial in *El Monitor Repúblicano*, the choices were to "exterminate

paskan languages by Johann Karl Eduard Buschmann (1805–1880) during the 1850s and 1860s.

60 Raymond Williams coined the phrase "selective tradition" to refer to "an intentionally selective version of a shaping past and a preshaped present, which is then powerfully operative in the process of social and cultural definition and identification." Williams, *Marxism and Literature*, 115. For a suggestive engagement with the relationship among sedentarism, the state, and history, see Deleuze and Guattari, *A Thousand Plateaus*, esp. chap. 12.

61 Pimentel, *La economía política*, 186.

62 Ignacio Ramírez, as quoted in Brading, *The First America*, 661.

them or civilize and mix them with other [races]."[63] It is perhaps easy to caricature such perspectives and overstate the degree to which mid-century thinkers denigrated the *indígenas* of their own age, but it is also clear from their own intellectual production that only ancient Indians were worthy of validation and recognition.

This is precisely the effect of the image: the archaeological emphasis glorified indigenous culture from a temporal distance, rendering Indians as inanimate objects of natural history and artifacts for contemplation.[64] The collapsed heads, the encroaching jungle, and the crumbling structures all suggest the past rather than the present, consciously drawing attention to the primordial while erasing its threat. The inclusion in the image of what appear to be tourists, at the far left and far right, underlines the point. They stand in profile or with their backs to the viewer, active participants in the scene. In sharp contrast, the two Indians included in the image are inert parts of the scene itself. Positioned in the foreground and posed, with spears in hand, as if in a diorama, they appear as romantic and proud warriors of a bygone era, impressive yet impotent. This subtle blending of archaeology and ethnography merely confirmed that the Indian had been reconstituted as an object of historical study rather than a subject with political will.[65] In other words, as Mary Louise Pratt has observed, indigenous history and culture had been "revived *as dead*," resuscitated with the fetid breath of archaeology.[66]

63 Quoted in Escalante Gonzalbo, *Ciudadanos imaginarios*, 57n8.

64 In this sense, scientific discourse—archaeology, history, and the developing field of anthropology—reconfirmed the distinction between self and other that had previously been mediated by a colonial legal system that recognized two independent republics, *la república de españoles* and *la república de indios*. See González Echevarría, *Myth and Archive*.

65 On the conflation of archaeology and ethnography, see Wolfe, "History and Imperialism," 410.

66 Pratt, *Imperial Eyes*, 134. Similarly, Luis Villoro notes that the Indian at mid-century was "no longer present" but rather reduced to a "pretty archaeological theme." Villoro, *Los grandes momentos*, 161. Pratt refers specifically to the "European imagination" and how it produced contemporary non-European peoples as "archaeological subjects." Although not speaking of Europeans, I think the general concept translates quite well to late-nineteenth-century Mexico, in which a Europeanized imagination revived the indigenous past both for its own contemplation and for that of its European counterparts. See also Widdifield, *The Embodiment of the National*, esp. chap. 3.

FIGURE 3. The chronotopic landscape: Detail from García Cubas, *Carta general.* Courtesy of the Mapoteca Manuel Orozco y Berra, Mexico City. Photograph by Carmen H. Piña.

García Cubas complemented this cultural landscape with a natural one in the other pictorial image on the map (figure 3). In contrast to a forbidding and overgrown land threatening archaeological ruins with absorption, here nature appears almost manicured in its perfection. We see seven prominent topographical features—Organos de Actopan, Iztaccihuatl, Cofre de Perote, Popocatepetl, Montañas de Jacal, the Pico de Orizaba, and the Cascada de Regla—organized as a panorama of the beauty of the Mexican landscape. This was, it should be emphasized, anything but a "natural" landscape. It was a cultural construction, a way of seeing, and an active ordering such that *landscape* takes on both its verbal and nominal meanings.[67] In fact, it was an impossible scene in which García Cubas collapsed features from spatially disparate areas into one frame and arranged them for theatrical force: the images to the far left and far right serve as curtains pulled back, drawing the viewer's eyes to the centerpieces of the performance, the idealized and snow-capped peaks. A simultaneity of perspective is given in which the viewer sees the Pico of Orizaba and the peaks of Popocatepetl and Iztaccihuatl

67 On landscape as a way of seeing, see the critical perspectives of Cosgrove, "Prospect, Perspective and the Landscape Idea," and Barthes, *Mythologies,* esp. 74–77. A more celebratory view can be found in Schama, *Landscape and Memory.* For varying perspectives on the separation of "nature" and "culture," see Glacken, *Traces on the Rhodian Shore;* Smith, *Uneven Development,* esp. chap. 1; and Coronil, *The Magical State,* part 1. "Way of seeing" comes from John Berger's classic *Ways of Seeing.* On understanding landscape as both verb and noun in the geographic ordering of the world, see Barnes and Gregory, eds., *Reading Human Geography,* 292.

from different locations. The land appears unusually ordered and organized; any signs of human presence have been removed. Unlike the previous cartouche, here the native presence is obvious only through its absence. Mexico appears as a land of sublime contemplation, where one could "admire the sublimity of nature," a trope García Cubas and others inherited from Alexander von Humboldt's own views of Mexico and its landscape.[68]

There are, of course, multiple ways to interpret the landscape image. One could argue perhaps that the image helped promote Mexico as a pleasing prospect for potential colonizers and investors, rendering Mexico familiar to foreign viewers and as an untouched arcadia of vast proportions, replete with open land for cultivation and water for irrigation. In an era of increasing calls for attracting (white) immigrants to settle in underpopulated regions of the country, calls that would echo most loudly in the coming decade in the works of men such as Francisco Pimentel and García Cubas, the image's visual harmony might have helped further Mexico's image as "one of the choicest countries in the world for colonization."[69]

But perhaps the image functions on a more subtle level. The image is in fact not a panorama of the Mexican landscape but a portrait primarily of features from a sharply delimited part of the country: the corridor that connects the port of Veracruz to Mexico City. Most travelers to Mexico—be they diplomats, artists, journalists, scientists, or military personnel—reached Mexico City through this corridor, at least until the mid-1880s when railroads connected the northern border town of El Paso to Mexico City. These were, then, the topographical features with which most of the European and U.S. reading public would be familiar, features that had long played a leading role in the topographical imaginary of what Mexico "looked like." Repeatedly replicated in the travel literature, artistic portraits, trade journals, and military narratives of the period, the promontories of Popocatepetl, Iztaccihuatl, and the Pico of Orizaba were metonyms for Mexico itself.[70]

And more than metonyms. They were also, to use Bakhtin's term,

68 Mayer, *Mexico*, 1:1. A useful intellectual history of the idea of nature in Mexican history is Ruedas de la Serna, *Los orígenes de la visión paradisiaca*.

69 García Cubas, *The Republic of México*, 6; Pimentel *La economía política*.

70 On metonymy, see White, *Metahistory*, introduction. On metonymy and geographical landmarks, see Burnett, *Masters of All They Surveyed*, 130.

chronotopes: historically charged "points in the geography of a community where time and space intersect and fuse."[71] Prior to all of the gringo armies, bourgeois travelers, and scientific expeditions, Cortés traversed this corridor, viewed these topographical features, and conquered an empire. The Spanish conquest invariably influenced the way new arrivals experienced and viewed their surroundings. In their accounts it seems as if every action, to some degree, was imbued with the weight of the conquest. Sometimes literally: visiting a bar in front of which stood a cart that "might have come over from Spain with Cortés," W. E. Carson described how, with twenty-five silver dollars in his pockets, he "felt like an ancient Spanish galleon loaded with pieces of eight."[72] But it was specifically Cortés's historical presence that weighed most heavily on the minds of travelers, and it was his march to Tenochtitlán, described in epic detail in the copies of William Prescott's *The Conquest of Mexico* that they brought with them, that most captured the imagination.[73] Traversing the "scene of Cortez's [*sic*] wondrous exploits," they imagined themselves "off to the land of the Aztecs," gazed at the "snow-capped summit of Orizava [*sic*] . . . and the lofty barrier of mountains, encompassing the land," and found themselves "anxious to scale their heights and penetrate beyond."[74] The physical space of central Veracruz thus assumed form as a theatrical stage space for many travelers who integrated the epic of Cortés's expedition into the drama of their own passages.[75] In effect, García Cubas's image served to conjure a particular historical trajectory through the use of a number of the most symbolically significant

71 Bakhtin, *The Dialogic Imagination*, 7.

72 Carson, *Mexico: The Wonderland of the South*, 7, 10.

73 Alfred Siemens notes that the two books that seem inevitably to have accompanied the nineteenth-century traveler to Mexico were Alexander von Humboldt's *Political Essay on the Kingdom of New Spain* (1811) and William Prescott's *The Conquest of Mexico* (1843). Siemens, *Between the Summit and the Sea*, 53.

74 Shepard, *The Land of the Aztecs*, 11–12. See also Mason, *Pictures of Life in Mexico*, and Gray, *Mexico as It Is*. The rage for retracing Cortés's footsteps continued: see, for example, Franck, *Trailing Cortez through Mexico*, who stated his purpose was to "trace in its entirety the route followed by Cortez [*sic*] in his Conquest of Mexico," which he suggested had never "been correctly done either by historians or by modern travelers" (vii).

75 A good selection of travel accounts through Veracruz is available in Poblett Miranda, ed., *Cien viajeros en Veracruz*. See also Siemens, *Between the Summit and the Sea*.

geographical features in the nation's official history. Cortés's contingent and precarious journey became a fixed itinerary, a set piece of passage that functioned as the primary trope for both imagining and entering Mexico.[76]

With García Cubas's map, and its enframing of the *land* as a quintessentially Mexican *landscape* of sun-tinged Orizaba and beckoning Popocatepetl, the viewer, not just the traveler, could now engage in a symbolic reconquest of Mexico, traversing the land in stride with the *conquistadores* from the comfort of an armchair. As we will see in the next section, intellectuals at the SMGE were eager to fix Cortés's passage on the map with more than vistas of the countryside.

IV. NARRATION

In 1860, two years after the publication of García Cubas's atlas, the readers of the *Boletín de la Sociedad Mexicana de Geografía y Estadística* encountered a fervent, if brief, essay by one of its members, José Guadalupe Romero.[77] Entitled "Dictamen sobre los inconvenientes de mudar los nombres geográficos de las poblaciones de la República, aprobado por la Sociedad," Romero's article eschewed description and an apolitical pretense. Rather, this was an unabashedly prescriptive essay intended ultimately for the eyes of the president of the Republic. Romero's essay would have remarkable resilience and resonance over the course of the next half-century as members of the Sociedad and state officials repeatedly referred back to it as a founding statement on the politics of place-names.[78]

The report stemmed from an investigation, headed by Romero, charged with "selecting the best means to avoid the disorder and confusion caused to our geography, and even the science of geography in

76 On the use of a small number of historical itineraries in elaborating a quintessential image of the colony or nation, see Cohn, *Colonialism and Its Forms of Knowledge*, and Poole, "Landscape and the Imperial Subject."

77 Romero, "Dictamen sobre los inconvenientes."

78 For example, see García Cubas, Díaz Covarrubias, and Fernández, "Dictamen sobre los inconvenientes de variar los nombres"; Díaz, *Memoria de la Comisión Geográfico-Exploradora ... 1878 a 1879*, 26–27; and García Granados et al., "Dictamen presentado a la Sociedad por los socios ingenieros Ricardo García Granados, Francisco de P. Piña y Ricardo Ortega y Pérez Gallardo."

general, by the changing of the names" of cities and towns in the country. Perhaps perceiving the volatility of naming as an allegory for Mexico's own political instability, Romero turned geographical inconvenience into ominous portents for political society, alluding repeatedly to the "serious problems," "ills," and "disorders" occasioned by such changes. The nation would face "serious problems," he warned, if the changing of geographic place-names continued unabated. Indeed, "civilized nations," he wrote, "had been very cautious" with regards to the changing of place-names. The span of centuries and waves of political crises had done little to alter the names of "Memphis and Thebes, Babylon and Ecbatana, Jerusalem and Athens," geographic names that had far outlived those that named them. Even in the "heat of revolution," he argued, the French had only changed the name of one place, the capital of the department of Vendee, and even this single variation had given rise to interminable problems. Thus, he concluded, certain issues require that the national government have exclusive rights to pass laws and "who can doubt that among this class of laws are those that fix invariably the Geographical names of the nation?" The committee under Romero's direction offered a solution: proclaim that only the federal government, rather than the federated states or municipal authorities, would have the power to change geographic place-names.[79]

In an era of civil war between centralists and federalists these were sharp words. And in fact, the government, under the federalist leadership of Benito Juárez, rejected Romero's proposal, most likely because of concerns over alienating municipal and state leaders whose support was critical in the aftermath of a three-year civil war and in the midst of a foreign invasion. A memorandum sent to all the state governors at the orders of Juárez stated that, while the government remained "convinced of the inconveniences caused by the frequent changes of names," it did not believe it "necessary to force State authorities to prevent such changes." Rather, it would be enough to "make them aware of the need to support the petition of the Sociedad."[80] But what compelled Romero to make such an impassioned plea for something as seemingly innocu-

79 Romero, "Dictamen sobre los inconvenientes."

80 Circular 160: Ministerio de Justicia, Fomento e Instrucción Pública, Sección de Fomento, May 13, 1862. Reprinted in BSMGE, 1a. época, tomo 8 (1860), 448–49.

ous as onomastic permanence in the first place? Why was federal intervention necessary? And what were these ills and disorders that threatened to undermine the nation?

On one level, Romero's concerns were pragmatic. As he remarked,
if the arbitrary changing of place-names persisted, domestic rule and
international relations would be thrown into disarray and the mass of
confusion which already plagued "the charts, maps, statistical studies,
laws, and government orders" would continue. While alluding to the
administrative chaos that resulted from an uncodified landscape, Romero avoided any further explicit discussion along these lines. It was
left to his associates in the SMGE, among them García Cubas, to state
unequivocally a decade later that the constant and arbitrary changing of place-names by municipal and regional authorities impeded
the rationalization of taxation and undermined the foundations of
property.[81]

But Romero's concern over the changing of place-names was primarily an historical one. Permanent place-names, he observed, preserved
history by imprinting the land with a genealogy and he approvingly
cited German linguist Karl Bauschmann's observation that "Geographic place-names, by their permanence and duration, can be considered as precious monuments of remote times."[82] In an analogy revealing for its transparent overtures to state control, Romero compared
geographic place-names to the proper names given by the state to its
citizens. Arguing that proper names had "always been understood as
immutable [. . . and] as the only means of identifying people and dis-
tinguishing among them," he suggested that only in the rarest of cases
should an individual be permitted to change their name, and then only
through addition of a new surname on to their old one.[83] Similarly, how
could Mexico's own deep genealogy be substantiated if place-names
were not permanent? Romero continued his analogy: "With even more

81 García Cubas et al., "Dictamen sobre los inconvenientes de variar los
nombres."

82 See also the work of Antonio Peñafiel, who argued that place-names
preserved tradition in places where "history has completely disappeared." Peñafiel, *Nomenclatura geográfica de México*, vi; and Peñafiel, *Nombres geográficos de
México*.

83 Romero, "Dictamen sobre los inconvenientes." On the relationship between permanent names and state control, see Scott, *Seeing Like a State*, chap. 2.

reason Geographic names of places should enjoy such immutability and duration, as men, in the end, exist for few years, and frequently change their situation and disappear; but places are permanent and firm, destined to contain races and generations, and to be mute witnesses of history."[84] Place-names, particularly indigenous place-names, were mute witnesses that spoke volumes: they spoke *through* their names and constituted the means by which a history of "races and generations" could be reconstituted.

Such concern with indigenous place-names was part of a developing indigemanía and constituted a prime component of the more general trend described by Edward Said as the postcolonial "search for authenticity."[85] Unlike the colonial project, in which naming (or more correctly, renaming) functioned as a routine mechanism for possession, in which a new cultural presence was imprinted onto the land to both confirm and create a space upon which colonization could occur, the nationalist project resurrected or actively perpetuated names that alluded to a history prior to the colonial origin.[86] Never did a Mexican official, when confronted with a confusing administrative landscape, suggest that the names of smaller villages and towns be replaced by their global

84 Romero, "Dictamen sobre los inconvenientes."

85 Said, *Culture and Imperialism*, 226. To what degree nineteenth-century Mexico can be described as "postcolonial" is open to question. The term itself, for all (or because of) its prominence in academic discourse, is susceptible to a number of interpretations and usages. As the point of this chapter is in part to draw attention to the discursive devices used to simultaneously efface and confirm persistent forms of structural inequality and domination, "postcolonial" in this instance should not be understood as a description of some entirely new situation that superseded colonialism. As Arif Dirlik observes, such an understanding "mystifies both politically and methodologically a situation that represents not the abolition but the reconfiguration of earlier forms of domination." Dirlik, *The Postcolonial Aura*, 54. José Jorge Klor de Alva questions to what degree one can even refer to the *gente de razón* who assumed power in 1824 as previously "colonial subjects." See Klor de Alva, "The Postcolonization of the (Latin) American Experience." See also the cautionary remarks on colonial and postcolonial discourse in Adorno, "Reconsidering Colonial Discourse."

86 The literature on colonialism, naming, and possession is extensive. For a sampling, see Boelhower, "Inventing America"; Clarke, "Taking Possession"; and Seed, *Ceremonies of Possession*. For an analysis of naming practices that goes beyond the commonplace assertion that naming constituted a form of cultural possession, see Carter, *The Road to Botany Bay*.

coordinates of longitude and latitude, as one colonial official suggested be done in British India.[87] The names themselves were living sources of that deep history so critical to contemporary geographic and historical self-fashioning. When García Cubas resurrected Romero's article in 1869, he observed that indigenous place-names often signified an idea, recorded a historical event, or indicated the topographic situation of a place.[88] Such names were a critical dimension of his own endeavors, begun in 1857 and eventually published in 1885, to reconstruct the historical route taken by the Mexica from the northern reaches to the central plateau.

Obsessions with the pre-Hispanic history never went so far as to completely obscure Mexico's pretension to European consanguinity. Indicative of the tension over history, José Guadalupe Romero had another route in mind and his genealogical analogy had a second meaning. The catalyst behind the formation of the committee that he directed stemmed from the inability of SMGE cartographers to trace Cortés's route from Veracruz to Mexico City in the first general map of the Republic. The map, while primarily intended as a symbolic affirmation of the country's "political respectability" after the crushing loss of half the national territory in the war with the United States, was also designed to construct a visual narrative of Mexico's "*true* history."[89] The cartographers who composed the map had been assigned the task of retracing and marking "with exactitude" two historical routes considered foundational journeys in Mexico's own path to self-realization: Agustín de Iturbide's itinerary from Iguala (where he delivered his famous proclamation) to Mexico City (where the Spanish capitulated) and Cortés's route from the coast to the Aztec capital of Tenochtitlán. After careful study, the researchers reconstructed Iturbide's route; that of Cortés proved more elusive.

Cortés's second letter to Emperor Charles V, written at some point during the autumn of 1520, served as the primary source for the cartographers. In the letter, Cortés narrated his journey from his landing on the gulf coast to Tenochtitlán:

87 Edney, *Mapping an Empire*, 115.

88 See García Cubas et al., "Dictamen sobre los inconvenientes de variar los nombres," and Buelna, "Peregrinación de los Aztecas y nombres geográficos indígenas de Sinaloa."

89 *Boletín de Geografía y Estadística de la República Mexicana presentado al Su-*

I traveled for three days through the country and kingdom of Cempoal. . . . [O]n the fourth day I entered a province which is called Sienchimalem, in which there is a town which is very strong and built in a defensible position on the side of a very steep mountain. . . . Then I went over a pass which is at the frontier of this province, and we called it Nombre de Dios, because it was the first we had crossed in these lands [. . . and] on the slopes below the pass there are other villages and a fortress called Ceyxnacan which also belongs to Mutezuma. . . . From there I continued for three days through desert country which is uninhabitable because of its infertility and lack of water and because of the extreme cold. . . . After three days we crossed another pass. . . . [W]e called it Firewood Pass. On the descent from this pass . . . there is a valley thickly inhabited. . . . [T]his valley and town are called Caltanmí. . . . After staying there four or five days, I left them all very pleased and went up the valley to the town of the other chief I spoke of, which is called Ystacmastitán.[90]

These brief references were most likely supplemented by those found in Bernal Díaz del Castillo's *True History of the Conquest of New Spain.* They constituted sparse references indeed for reconstructing an historical itinerary. Even more problematic was the fact that the indigenous place-names mentioned in the texts had seemingly vanished from the landscape, victims of repeated name changes and presumably lost forever. As Romero observed, "in the end, two places along the Spanish conquistador's route remained in doubt, because one could not definitively determine what the names today are of places that prior to the conquest had different names."[91] Romero did not state which two places these were, nor did he provide any explicit indication as to the nature of the name changes. García Cubas, however, did: since the 1830s indigenous place-names were being supplanted by those of Mexican patriots of Independence—Allende, Morelos, and Hidalgo, among others. Place-names had also become weapons in the political wars between competing factions vying for local, regional, and national control. Little wonder that García Cubas lamented that "there are States that have

---

premo Gobierno por la Comisión de Estadística Militar 3 (August 1849), quoted in Mendoza Vargas, "Historia de la geografía," 73. My emphasis.

90 Cortés, *Letters from Mexico*, 54–57.

91 Romero, "Dictamen sobre los inconvenientes."

changed nearly all the names of their towns."[92] Such concerns led him to argue that while "perpetuating the memory of illustrious men" was a laudable goal, a statue or a monument would better achieve such an end.[93] Just as problematic, places often had multiple names: one town could have simultaneously an ecclesiastical, indigenous, common [vulgar], and legal name, all of which were substantially different and not necessarily categorically fixed and archived. By the century's end, a new state exploration agency would have as one of its primary tasks the collection and cataloguing of all such names in order to overcome such ambiguities.[94]

But at mid-century finances did not permit quick resolutions. The SMGE cartographers never recreated what was thought to be Cortés's precise route, and they excluded the itinerary from the final map.[95] It was this perceived failure that provoked a strong response, at least among the members of the SMGE, and Romero's subsequent letter.[96] They understood the power of foundational narratives as a means to geohistorical legitimacy. For Romero, like his intellectual predecessors José María Luis Mora and Lucas Alamán, and his contemporaries Francisco Pimentel and Manuel Orozco y Berra, Mexico's genesis began with Cortés and his epic confrontation with the Aztecs. If Cortés's travels and travails were acts of national foundation, how could they be contingent and ambiguous? If the nation's roots were to be found in Cortés's route,

92 García Cubas, *Memoria para servir a la carta general*, vi.

93 See García Cubas et al., "Dictamen sobre los inconvenientes de variar los nombres," and García Cubas, *Memoria para servir a la carta general*.

94 See Díaz, *Memoria de la Comisión Geográfico-Exploradora . . . 1878 a 1879*, and Díaz, *Exposición internacional colombina de Chicago en 1893*, 16. I examine the work of the Comisión Geográfico-Exploradora in more detail in chapters 4 and 5.

95 Years later Manuel Orozco y Berra would suggest that the route was also not included in the map due to scale, but I have found no evidence that such an issue was originally raised.

96 Romero was not merely a quirky crank but representative of the period's obsession with attaching history to territory. Not only were Romero's arguments reiterated by García Cubas and others, but the first president of the SMGE, José María Justo Gómez de la Cortina, devoted years of his life to the creation of a *Diccionario de voces necesarias para el estudio de la cosmografía, geografía y topografía para la inteligencia de las relaciones históricas y de viajes*, which was never completed or published.

how could it be anything other than a solid, firm line, boldly coursing across the center of the page?

All nation-states have their founding myths, overdetermined and evolutionary narratives of the nation-state's history that lend credence to certain claims to rule. Such myths are genealogical teleologies that give power to the presumed legitimacy of descent. Not only are these foundational myths a mode of historical narration, they are often the purpose (and politics) of history itself, a hallowed means of imposing a temporal order—a chronology—upon the contingencies and multiplicity of what has come before in order to tell a coherent, delimited, encoded story.[97] Nationalist history, however, has little power without a space upon which to unfold. As D. W. Meinig has wryly put it, "[H]istory takes place."[98] A good deal of the fascination with geography, and the inscribed national map, revolved around the understanding that it could create those places upon which a foundational, sequential history could be narrated; upon which one could "follow the march of armies, [or] the routes and discoveries of travelers" and compose the course of history itself.[99]

The geographical concerns of Sociedad intellectuals such as Romero and García Cubas were thus simultaneously historical ones: the possibility that toponymic turmoil would make Mexico's own historical genealogy unstable, interrupting the "narration of the nation," and with it, the enterprise of history itself.[100] By putting a stop to the arbitrary changing of place-names, historical as well as geographical ambiguity would be reconciled through a form of spatial order and cartographic permanence. Only then could a retrospective coherence be constituted and a destiny deciphered, in the process drowning out the cacophony of competing histories of the land.

97 See Carter, *The Lie of the Land*, 211. Useful studies that elaborate on this theme include Carter, *The Road to Botany Bay*; Obeyesekere, *The Apotheosis of Captain Cook*; Duara, *Rescuing History from the Nation*, esp. chap. 1; and Trouillot, *Silencing the Past*. On historical narration and "emplotment," see White, *The Content of the Form*; Ricouer, *Time and Narrative*; and Somers, "The Narrative Constitution of Identity."

98 Meinig, "The Continuous Shaping of America," 1205. See also Pred, "Place as Historically Contingent Process."

99 Orozco y Berra, *Materiales para una cartografía mexicana*, vii.

100 The phrase is Homi Bhabha's. See Bhabha, ed., *Nation and Narration*.

CONCLUSION

In a panegyric to García Cubas, delivered to the members of the Sociedad in 1909, Francisco de P. Piña praised history and geography as two "beautiful sciences powerfully united," which, together, provided a "real and positive, rather than ideal and platonic, conception of the nation."[101] Together they explained not only "the reason for its existence" but also the rationale behind "the sacrifices which arise from the defense of its inviolable rights of sovereignty and independence." History and geography were more than beautiful sciences: they were national sciences, two sides of a single complex that gave a territorial expression its historical legitimacy and history a territory upon which to unfold. Together, they wrote a narrative around which modern Mexico could cohere.

The praise for García Cubas is not surprising. His 1858 carta general—as much biography as geography—served in many ways as a defining moment in Mexican nation-state formation. Nationhood itself found expression in the disciplinary unification of history and geography on the surface of his national map, which, through a subtle blending of technology and iconography, portrayed Mexico as both timeless yet richly historical. The scientific naturalization of the nation's territory, the visual production of a particularly Mexican cultural and natural landscape, and the narration of its presumed roots converged on his map to fix the land as a stable cultural, political, historical, and geographical object.

In this sense, García Cubas's carta general has more in common with the indigenous maps of preconquest and early colonial Mexico than might at first seem apparent. Barbara Mundy's careful examination of the cartographic images produced as part of the *Relaciones Geográficas* for the Spanish monarch in the late sixteenth century revealed how a particular genealogy and history was woven into the maps of indigenous groups whom the Spanish conquered.[102] While the surface differences between these pinturas and a map such as García Cubas's are paramount, there is at least one similarity worth noting: both sets of images func-

101 De P. Piña, "Importancia de los trabajos," 390.
102 Mundy, *The Mapping of New Spain*, esp. chap. 5.

tion to bring history and geography into mutual dialogue in order to legitimate the position of those in power in relation to those over whom they presume to rule. By analyzing the internal form and content of the map as well as the historical circumstances that conditioned its production, I have suggested that the presumably objective and "modern" scale maps that appear so natural to us are themselves powerful stories about the past and the present, replete with their own ideological presuppositions.

The insertion of a history into the map can lead, ironically enough, to anachronistic history. As Thongchai Winichakul has observed in his careful analysis of the cartographic creation of Thailand, by taking as a given the prior existence of the Thai nation-state, historians miss the fact that it was in the process of being created, and at the expense of a whole array of smaller, dispersed kingdoms. They thus inadvertently replicate and give credence to a nationalist discourse that arose at a specific point in time to legitimate certain spatial claims to power.[103] Timothy Anna's study of nineteenth-century Mexico makes a similar, although no way near as comprehensive or daring, claim: in assuming the existence of a coherent entity called Mexico, "uninterrupted, since the time of the ancient civilizations," historians have codified the legitimation discourse of late-nineteenth-century elites who constructed the early years of independence as a moment of threat to an already constituted space, thus reducing the history of Mexico to a metropolitan-

---

103 Thongchai, *Siam Mapped*, 146–48. Part of the power of Thongchai's critique is the way in which he is able to connect such histories with broader trends in contemporary historiography and politics. Historians take the existence of Thailand as a given because they feel compelled to see the history of Thailand primarily through the lens of international relations and colonialism. Thus an already integrated nation-state (Thailand) confronts the Western powers. Such an epistemological grid excludes the "voice of those tiny states which were never born later as nations, despite their active role at that time, and allows only the story of the emerging nation to be heard" (147). It also excludes from consideration the manner by which the Bangkok elite appropriated Western mapping technologies to assert control over those locales. Thongchai's assertions offer comparative food for thought for nineteenth-century Mexicanists. José Jorge Klor de Alva's questioning of the applicability of the term "postcolonial" to Latin America already constitutes one such fruitful comparative engagement. See Klor de Alva, "The Postcolonization of the (Latin) American Experience."

narrated nationalist tragedy of territorial aggregation and disintegration.[104]

A historical and spatial domain called Mexico was never external and prior to history, like coordinated points in the graticule. It was actively constituted, and narrated, and national maps played a foundational role in that process. This should not be construed as suggesting that García Cubas's image was an alien representation that obscured or erased a more "authentic Mexico." Precisely the opposite: there was no "authentic" landscape or essential Mexico hiding behind the facade of the image, waiting to emerge, pristine and untouched, from the primordial mists.[105] It was precisely the search for such a fixed essence that proved so critical to defining the state-in-formation. In this sense, García Cubas's carta general was the broadest of a whole array of what I will be calling "state fixations": federal obsessions with the permanence—the fixity—perceived as basic to the practice and theory of government and to state-promoted capitalist development. Rule and development required the capacity to count, to tax, to arbitrate—in a word, to regulate. Regulation and investment in turn required homogenous, predictable space and fixed units of analysis: political jurisdictions, categories of people, bounded properties, and so forth. But, as Mariano Otero's lamentations (with which this chapter began) clearly suggest, faced with the imperialist pretensions of a northern neighbor, the most basic unit of analysis in need of "fixing" was the nation-state itself.[106] Hence the obsession with

104 Anna, *Forging Mexico*, 6. For a rich reconsideration of the first half-century of independence, see Escalante Gonzalbo, *Ciudadanos imaginarios*.

105 My critique of authenticity is not meant to suggest that we supplant such searches for a legitimating origin with an uncritical celebration of hybridity and pluralism. As Seamus Deane has trenchantly remarked, this simply substitutes "one kind of colonizing experience [with] another"—that is, that of multinational capital. Rather, the point is to understand how an "authentic" point of origin was created and visually articulated, for both domestic and foreign elite audiences, at a particular point in time to legitimate the exercise of power. See Deane, introduction to *Nationalism, Colonialism and Literature*, by Eagleton, Jameson, and Said, 19. See also Mitchell, "Different Diasporas and the Hype of Hybridity." On questions of authenticity and representation, see O'Hanlon and Washbrook, "Histories in Transition." See also the exchange between O'Hanlon and Washbrook and Gyan Prakash: Prakash, "Writing Post-Orientalist Histories of the Third World"; O'Hanlon and Washbrook, "After Orientalism"; and Prakash, "Can the Subaltern Ride?"

106 On how the idea of the state has to be constructed and sustained, see

a very material fixation: the carta general. Hence also the debates over permanent place-names, the fetishism for sedentary—fixed—Indian civilizations, and the reification and naturalization of highly contestable and contingent versions of history and space into official portraits of the past. The subtle connection of history and geography offered an ideologically saturated and finished image of a nation-state in formation, in which a developing terrain appeared as tradition and a multiplicity of spaces were reduced to the linearity of a singular narrative.[107] Such erasures left only one story to be told: that of the state itself.

Abrams, "Notes on the Difficulty of Studying the State"; Corrigan and Sayer, *The Great Arch;* and Sayer, "Everyday Forms of State Formation."

107 See Bhabha, "DissemiNation," esp. 148–50.

CHAPTER 2

Fugitive Landscapes

On a clear day in 1874 Antonio García Cubas ascended the hill of Macuiltepec on the outskirts of Xalapa. From its commanding heights, he contemplated the dramatic landscape of central Veracruz. Facing south, he gazed out across a vast expanse of semitropical greenery presided over by the dormant Cofre de Perote and punctuated by the church towers of Teocelo, Coatepec, and Xico. His eyes were drawn to the rocky outcrop of the sierra of Huatusco, which faded into the "whitish blue" of the hazy horizon until, far off in the distance, the snow-capped symmetry of the Pico de Orizaba reestablished the ground and refocused his view. Turning north, the serrated lattice of the sierra of Chiconquiaco dominated the vista. A less pleasing view, he mused, but spectacular nonetheless, particularly when the sierra's rough edges and ledges, "crowned" by the town of Naolinco, sparkled under the sunlight in brilliant contrast to the shadowed abyss of the canyon of Actopan.[1]

For all its grandeur, this "most picturesque of landscapes" was also a relatively uncharted and, at least to metropolitan travelers such as García Cubas and even *xalapeño* officials, unknown one.[2] Few large-scale topographic or cadastral maps had been made of the area; no exploration commissions or survey teams had traversed the terrain; and no systematic attempts had been made to gather and order local documents, land titles, and other information residing in municipal town halls. A grand *view* did not equal a concise *overview*, and it was the latter that proved to be of primary importance to Veracruz officials, particularly in the sec-

1 See García Cubas, *El libro de mis recuerdos*, 612.
2 Ibid.

ond half of the nineteenth century when they attempted to implement and enforce the *repartimiento de terrenos comunales* (division of communal lands).[3]

The general parameters of this land division are well known: codified in the Lerdo Law of June 25, 1856, it constituted the means by which land worked in usufruct by a largely indigenous peasantry and owned collectively by municipalities would be transformed into individual freehold plots. Over the course of the next half century, land division became an ideological obsession—a fixation—among Mexican liberals, who saw it as the solution to a host of social, economic, and political problems. Socially, it would convert each Indian into a small landowner and would theoretically lead to the "easy and spontaneous regeneration of the race"; economically, it would facilitate the creation of a property cadaster and the "benign valuation" of property taxes; and politically, it would resolve thorny questions over municipal, cantonal, and state limits through the creation of authoritative maps.[4] While these social and economic "problems" have attracted scholarly interest, the political issue of land boundaries and their cartographic certainty has been left largely unexamined. Yet issues of boundaries and the need for reliable maps, often seen as merely a technical backdrop, were critical to the land division. The state required that community boundaries first be determined and fixed [*fijado*] before lands were divided, making the land division contingent upon the ability to determine and map limits. Thus the division of communal lands can be understood as both a metaphorical and literal "state fixation": an obsession with land privatization based upon classical liberal economic assumptions and a recognition of the concomitant need to codify the landscape as a permanent, mapped, and fixed entity.

This was easier said than done. If the *sierra veracruzana* was not a fully known quantity—explored, mapped, and fixed—neither was it a blank slate awaiting topographic inscription. As bureaucrats and officials knew all too well, it was an inhabited and working land dense with

3 On the relationship between "views" and "overviews," see Burnett, *Masters of All They Surveyed*, 167.

4 Circular 9, Sección de Gobierno, March 9, 1886, CLEV—*1886*, 8–16. The author included a fourth benefit to be derived from the land division: the "moral glory" Veracruz would attain in having been the first to "concern itself with the knowledge and division of property."

meaning and history, woven together with the social fabric of experience and practice, and riven by class and community conflict. As a consequence, measurement competed with memory, inscription with inheritance, and technical abstraction with social experience as arbiters of reality. In trying to simplify and codify a landscape of overlapping jurisdictions and use rights, of ambiguous borders and shifting place-names, state officials had to reconcile a profusion of contradictory and competing claims with the few remnants of documents available in municipal archives. At those moments, the land must have appeared as more than merely unknown: it must have seemed almost fugitive, as if it were an accomplice in a larger conspiracy undermining their efforts.

I. BORDERS

In late-nineteenth-century Veracruz, state officials obsessed over borders. One cannot open a state governor's *memoria* (report) without encountering repeated references to the need to fix [*fijar*] boundaries. The reason is understandable. Ambiguously defined village, municipal, and state limits hindered the rationalization of property taxation and raised repeated jurisdictional questions over land claimed by two (or more) municipalities.[5] Even cantonal and state limits fell prey to such problems over the course of the century, spawning at times acrimonious correspondence between different state officials.[6] A cursory examina-

5 *Memoria leida por el C. Gobernador del Estado ante la H. Legislatura del Mismo, el día 13 de Octubre de 1871*, 37. See also Circular 24, June 17, 1889, in *Memoria presentada a la H. Legislatura del Estado Libre y Soberano de Veracruz Llave, el 18 de Septiembre de 1890*. Note: the spelling of Xalapa varied over the nineteenth century between Xalapa and Jalapa. In the text, I use Xalapa, its modern spelling. In the notes, I use the spelling chosen by the author. As we will see in chapter 4, the spelling of the name was a subject of heated debate.

6 *Memoria leida . . . el día 13 de Octubre de 1871*, 37. For persistent problems over state limits and village boundaries, see, for example, Miguel Cid y León to Porfirio Díaz, November 21, 1889, CPD, leg. 14, caja 23, doc. 11480; Porfirio Díaz to Francisco Arce, Gobernador del Estado de Guerrero, April 18, 1888, CPD, leg. 13, caja 7, doc. 3223; and "El día cuatro de Enero de mil novecientos, reunidos en la Sala Municipal," AGEV, RG, Tierras, Reparto, 1903, tomo 2, exp. Terrenos: Lo relativo al reparto de los del Municipio de Ayahualulco. On cantonal limits see documents in AGEV, RG, Tierras, Quejas, 1880–1883, exp. Terrenos: Los vecinos de San José Miahuatlan del Cantón de Jalapa se quejaron de que la Muni-

tion of the self-described limits of municipalities submitted to a federal mapping agency by various mayors of the canton of Chicontepec in 1883 reveals that nearly every municipality in the canton that bordered the states of Hidalgo or Puebla was in dispute with pueblos in those states over borders, some of those disputes being quite violent.[7] At times multiple political units were layered into one confusing mass conflict, such as a lengthy dispute in the canton of Coatepec, which simultaneously provoked conflicts between the municipalities of Ixhuacan and Ayahualulco; the *cabecera* (district capital or seat) of Ayahualulco and the *sujeto* (dependent village or hamlet) of Soteapam; Ixhuacan, Ayahualulco, and the neighboring hacienda of Tenextepec; the cantons of Coatepec and Jalacingo; and the states of Veracruz and Puebla.[8] To complicate matters further, private survey companies [*compañías deslindadoras*], charged with surveying the so-called *terrenos baldíos* (vacant lands), took advantage of border ambiguities to increase the domain of their work and thus their payment in kind. Their opportunism caused persistent problems for both the federal and state administrations, intent upon demonstrating that such land surveys were not meant

---

cipalidad de Juchique de Ferrer, del Cantón de Misantla, pretende despojarlos de una parte de sus terrenos [hereafter Los vecinos de San José Miahuatlan]; and AGEV, RG, Tierras, Limites, 1899, exp. Límites: Lo relativo a los jurisdiccionales de los Municipios de Chiconquiaco del Cantón de Xalapa y Juchique y Yecuatla del de Misantla. Note: The lengthy citations in the notes for these documents is a reflection of the fact that they are still in the process of being organized and catalogued. I have attempted to provide as comprehensive a set of citations as possible for this documentation. In many instances foliation was either unavailable or more confusing than clarifying. I am extremely grateful to the staff of the Archivo General del Estado de Veracruz, who allowed me to consult these materials, and to Michael Ducey, who directed me to them.

7 ACGE, exp. 3, Cantón de Chicóntepec, October 2, 1883. On violence in these disputes, see Jefe Político del Cantón de Chicóntepec to Secretario del Superior Gobierno del Estado, June 17, 1879, AGEV, RG, Tierras, Límites, 1879–1892, exp. Límites. Lo relativo a la cuestión que existe entre Ylamatlan del Cantón de Chicóntepec y Coatitla del Distrito de Zacualtipam del Estado de Hidalgo, f. 13r–v and passim.

8 See AGEV, RG, Tierras, Límites, 1902, exps. Ixhuacan and Ayahualulco; AGEV, RG, Tierras, Reparto, 1903, tomos 1 and 2, exp. Terrenos: Lo relativo al reparto de los del Municipio de Ayahualulco; and documents in AGEV, RG, Tierras, Límites, exp. Ayahualulco e Ixhuacan, 1905–1914.

to "dispossess anyone but merely to determine what belonged to whom."[9]

Just as troublesome, border disputes impeded the division of communal lands. After all, how could a village's lands be surveyed and divided without first spatially defining the village itself? According to state law, land divisions could not proceed in villages with unresolved disputes over borders. Fixing the boundary line between villages thus became a procedural prerequisite for a land division.[10] In the sierra of Chiconquiaco in Veracruz, the difficulties in establishing boundaries ensured that the liberal fantasy of a fully surveyed and privatized rural property landscape would be perpetually frustrated. Border disputes were the most common reasons offered by village authorities to the state government for why the division of their communal lands had not proceeded apace. Delays lasted for decades as villages quarreled (or perhaps pretended to quarrel) over boundaries, and numerous conflicts and questions over land limits proliferated in the sierra in the latter half of the nineteenth century.[11]

A lengthy and contentious conflict between the villages of Tonayán

9 The accusation appears in Joaquín Alegre to Porfirio Díaz, October 10, 1886, CPD, leg. 11, caja 26, doc. 12650. The quoted text is from Carlos Pacheco to General Luis Mier y Terán, Gobernador del Estado de Oaxaca, April 17, 1885, CPD, leg. 10, caja 9, doc. 4333–35. On the *compañías deslindadoras*, see Holden, *Mexico and the Survey of Public Lands*.

10 The prohibition on land divisions in disputed regions originally appeared in Article 10, Decree 39, Sobre repartimiento de terrenos de indígenas y baldíos, December 22, 1826. See Florescano Mayet, "El proceso de destrucción de la propiedad"; and Ducey, "Liberal Theory and Peasant Practice." Article 12 of the state law on land division of the same year reaffirmed the state government's power to resolve the persistent questions and conflicts over borders, particularly those between communities in Veracruz and neighboring states, and emphasized cartographic clarity. See Trens, *Historia de Veracruz*, 6: 92–93.

11 To what degree villagers were pretending to quarrel to avoid land divisions is, of course, a complicated question and one that may be unanswerable. But the broader point of this chapter is to suggest that, whether intentional or not, that very ambiguity is what regional and federal officials wanted to bring to an end. The issue at hand was not to clear up who was really quarreling and who was not, but to transfer the power of arbitration and the maintenance of unambiguous knowledge into the hands of a certified bureaucratic agent beholden in some form to regional and federal authorities.

and Chapultepec serves as a case in point.[12] In December 1881 the authorities of Chapultepec sent a short letter to the prefect of Xalapa canton explaining why they had not divided their communal lands into individual plots in accordance with the law. The authorities of Tonayán, they explained, would not come to an agreement regarding their shared border, effectively inhibiting the division of the lands. Authorities in Tonayán countered by claiming they had already divided their lands in 1869 and that the border was not their responsibility but an issue to be taken up with the holders of the individual parcels.

The dispute between Tonayán and Chapultepec, which would persist over the course of the Porfiriato, was not exceptional in the sierra. A cursory examination of the general region of the sierra of Chiconquiaco, for one decade (1880–1890), shows the villagers of Tonayán simultaneously embroiled in conflicts with those from Chapultepec, Atexquilapam, Coacoatzintla, and Tlacolulan; Tlacolulan with Misantla and Altotonga; Juchique de Ferrer with San José Miahuatlan and Chiconquiaco, itself at odds with Yecuatla, San Juan Miahuatlan and Misantla.[13]

Conflicts among villages over liminal lands were not particularly new.[14] That they occurred in the sierra was understandable. A productive region, with fertile soils and abundant flora and fauna, a place where plums, cherries, and avocado were said to grow wild on slopes below extensive forests of elm and oak, the land was also extremely rugged and

12 AGEV, CLA, caja 2, exp. Tierras/Límites: Tonayán, 1878–1925 [hereafter exp. Tonayán].

13 See numerous documents related to these questions in exp. Tonayán; AGEV, RG, Tierras, Quejas, 1880–1883; AGEV, RG, Tierras, Límites, 1899; as well as in the reports of various prefects, such as those of Antonio Sánchez Esteva, "Memoria que rinde el Jefe Político del Cantón de Jalapa al C. Gobernador del Estado de Veracruz, 25 de Abril de 1887," and Teodoro G. y Lecuona, "Memoria que rinde el Jefe Político del Cantón de Xalapa al C. Gobernador del Estado de Veracruz, 28 de Mayo de 1895," both in García Morales and Velasco Toro, eds., *Memorias e Informes*, 3:29, 3:93.

14 Indeed, the AGN is filled with colonial documents describing in great detail conflicts between villages and between villages and haciendas. Although documentation for the early republican era is sparse, it is telling that in 1842 Mariano Galván Rivera compiled his *Ordenanzas de tierras y aguas* to help put property conflicts to rest. See Galván Rivera, *Ordenanzas de tierras y aguas*. For the colonial period, see Taylor, *Landlord and Peasant in Colonial Oaxaca*; Van Young, *Hacienda and Market in Eighteenth-Century Mexico*; Van Young, "Conflict and Solidarity in Indian Village Life"; and Borah, *Justice by Insurance*.

mountainous. Cross-hatched by a multitude of narrow valleys and imposing peaks, the sierra was a composite of microclimates. Elevation and temperature varied radically across relatively small distances. Villages less than a day's walk apart, such as Tonayán and Chapultepec, had extraordinarily different agricultural cycles and options.[15] Weather patterns could be extremely variable from one valley to the next, some of which only span a half kilometer at most. Ice and snow fell with regularity in portions of the sierra, such as Las Vigas and Chiconquiaco, while rarely touching others, such as Jilotepec and Coacoatzintla. While some villages suffered the cold winds that coursed from the nearby peak of the Cofre de Perote, others were spared. All of these climatic patterns wrought dramatic effects upon the cultivation of traditional crops such as maize, beans, and barley and affected the growth of particular kinds of vegetation used as fertilizer, of medicinal herbs and mushrooms, and of fruits from the caxixín and chaca trees.[16]

Just as important to cultivation were the so-called *nortes*, polar air masses that drop south in the winter, warming and picking up moisture from the Gulf of Mexico in the process. The precipitation they brought with them permitted agriculturists on the windward side of the sierra veracruzana to sow and harvest two, sometimes three, crops of maize and barley per year rather than the traditional one.[17] This area included the sierra of Chiconquiaco, where the nortes often shrouded the entire range in a thick fog. The crop of maize made possible by the nortes, sown in December and harvested in May or June, was referred to in the sierra as *tonalmil*, in contrast to maize planted during the traditional summer season, known as *temporal*.[18] Barley could also be planted and

15 The variation is so severe that when Joaquín María Rodríguez wrote his ecologic, ethnographic, and geographic account of the region, *Apuntes sobre el Cantón de Xalapa*, he chose to organize his discussion of villages in the sierra according to relative elevation.

16 The description is taken from María Rodríguez, *Apuntes sobre el Cantón de Xalapa*. Much of this general description still applies, but for temporal consistency I use the past tense only.

17 The assertion that cultivators could get three crops per year was made by British anthropologist Edward Tylor, who claims a local guide as the source of information, in his 1856 account of his excursion through Veracruz. See Tylor, *Anahuac*, 26. American traveler Frederick A. Ober confirmed Tylor's observations. See Ober, *Mexican Resources*, 14.

18 María Rodríguez, *Apuntes sobre el Cantón de Xalapa*, 11.

harvested twice a year, although not in the same land: barley planted in November or December was referred to as *aventurera*, while that planted in March or April was known as temporal.[19] Not all the inhabitants of the sierra reaped the benefits brought by the rains of the nortes equally however. Villagers at higher altitudes in the sierra, such as Tonayán (1,900 meters) and Chiconquiaco (2,300 meters), could not sow the tonalmil because the hardness of the winter earth impeded the quick growth of the plant. In its place, villagers grew potatoes, which they then exchanged for maize in regional markets. In sharp contrast, neighboring villages at lower altitudes, such as Chapultepec, Yecuatla, and Juchique, sowed the tonalmil with great success.[20] The quality of the temporal crop also varied according to elevation. Maize planted in lower altitudes required substantially less growing time and produced ears as much as five times larger than those produced in colder regions.[21] Little surprise then that many of the conflicts over borders were between villages at radically different elevations. Tonayán and Chiconquiaco would be repeatedly accused of invading neighboring lands at lower elevations.

Tonayán was particularly litigious and garnered a reputation as "stubborn" and "tenacious" in land disputes with neighboring villages such as Chapultepec and Atexquilapam.[22] Supposedly founded as a military outpost of the Mexican empire as a hedge against Totonac uprisings, Tonayán sat atop a rise in one of the most mountainous portions of the region. With a commanding view in all directions, the mesa served as a site for long-distance communication via fire signals with a fortress on the top of the hill of Macuiltepec (where García Cubas stood in 1874 taking in the panorama). But its strategic military value hindered its agricultural propensity. A limited zone of cultivation meant the land could rarely lie fallow, causing rapid soil degradation and forcing villagers to spend increasing amounts of time collecting fertilizer to meet their subsistence demands.[23]

---

19 Ibid., 12.

20 Ibid., 11–12. Notably, cultivators referred to the potatoes as their tonalmil crop.

21 Ibid., 181.

22 Ibid., 173.

23 Ibid., 165–75. The town name itself captures the sense of height and exposure: "where the sun can always be seen," or, as the current town motto puts it, "where the sun is born."

While important, ecological explanations do not completely account for the pervasive land disputes. Border disputes also flourished as the state tried to redefine and reorganize real property and achieve a simple but absolute synchronicity between agricultural property and political territory.[24] The provisions of the land division required that any territorial ambiguities—coincident use rights, "empty" spaces, zonal boundaries, or natural and imprecise border markers—be reconciled in order to give borders a vertical precision. To what degree villagers had lived with territorial ambiguities or in what manner they had negotiated certain kinds of tenure relationships is difficult to ascertain. Little is found in the documentation that would shed light on the issue; by their very nature such informal arrangements rarely appear in the archives. As E. P. Thompson observed, "Agrarian custom was never fact. It was ambience."[25] But some hints may nevertheless be discerned in the documentary record. For example, two villages in the canton of Coatepec —Ayahualulco and Ixhuacan—reached an agreement regarding their boundary line in order to clear the way for their respective land divisions. But carrying out the land division still proved to be problematic due to the status of a set of lands within the jurisdiction of Ixhuacan that had, for many years, been worked in usufruct by a number of villagers from Ayahualulco. Thus the authorities from the two villages inserted a clause into the agreement stating that the current occupants would continue to maintain usufruct rights to those lands and would be subject to the same fees as others in Ixhuacan. Such a situation proved untenable in the eyes of the state government, which required the clause be removed because it would complicate the land division.[26]

24 In one sense this already existed: as Cambrezy and Marchal have shown, in central Veracruz the boundaries of the hacienda, where it existed, tended to map directly onto cantonal boundaries. See Cambrezy and Marchal, *Crónicas de un territorio fraccionado*, 55–56.

25 See Thompson, "Custom, Law and Common Right," 102. A number of authors have pointed to the ways in which rights to land in the Huasteca and in portions of Morelos were overlapping and interconnected, relative rather than absolute. See Escobar Ohmstede and Gordillo, "Defensa o despojo?" and Hernández Chávez, *Anenecuilco*, 27–31. See also the suggestive comments raised by Eric Van Young in "Paisaje de ensueño con figuras y vallados."

26 "En el pueblo de Ixhuacan de los Reyes a los trece dias del mes de febrero de mil ochocientos ochenta"; and Jose Pichardo to Co. Gobernador, January 22, 1884, both in AGEV, RG, Tierras, Límites, 1902.

In the case of Tonayán and Chapultepec, what is remarkable is that neither village, in the initial years of the conflict, attempted to recount or describe the boundary between their respective lands. In their initial complaint, authorities of Chapultepec did not argue that the neighboring villagers of Tonayán had actually *invaded* their lands, but rather that they were unable to divide their lands because of the absence of an agreed-upon border. Similarly, the authorities of Tonayán did not claim that the lands in question were within the boundaries of their village per se but that, having divided the land into individual parcels, any question regarding the border should be addressed to owners of individual parcels rather than to them.[27] The absence of mention of a precisely fixed border in the villagers' own arguments did not persist as the case progressed, and by 1904 the villagers of Tonayán would claim that the border had always been La Magdalena River. But in the early 1880s it seems that, at least in terms of local practice, there was no permanently "fixed" and fastidiously defined border between the two villages.

I want to be clear that this is not to suggest that the ensuing conflict over limits was entirely new or that the villages had lived in harmony and peace up until that point. The tomes of historical documentation from the colonial era in the Archivo General de la Nación clearly demonstrate otherwise. Indeed, Tonayán's current Web site claims that villagers from Misantla founded Tonayán as an attempt to ward off land encroachments by villagers of Chapultepec![28] Rather, the point is twofold: first, sharp lines of political and proprietary demarcation are neither timeless nor natural.[29] In fact, and this is the second and more crucial point, the result of requiring villagers to precisely fix their borders could

27 See Jefe Político de Xalapa to Secretaría del Gobierno, September 11, 1882, exp. Tonayán, ff. 145r–149v; and Secretaría del Gobierno to Gobernador del Estado, February 23, 1884, exp. Tonayán, ff. 177r–81v.

28 http://www.cedem.ver.gob.mx/dirmun/cont/municipios/30187a.htm#

29 After independence village officials often requested an *apeo* or survey of their lands in order to have a legal document available in case of future problems with neighboring communities. These were clearly important efforts on the part of villagers to get territorial possessions confirmed by the new governing power. Yet such apeos did not map out clear, exclusionary, proprietarial rights, nor were they technical surveys of precise boundaries but rather kinds of ritual acts of possession, of walking out boundaries according to natural landmarks. For an excellent example, see the documents related to Los Reyes, Zongolica cantón, in ACGE, exp. 3.

often *spur* as many conflicts as it *resolved.* This was not unusual: William Taylor showed for colonial Oaxaca that "endless disputes" between Indian communities arose precisely when colonial officials tried "to make boundaries *more precise* [or] establish a *definite boundary line.*"[30] Thus late-nineteenth-century disputes over boundaries were not necessarily, or always, simple manifestations of age-old antagonisms that could only be resolved by the formalization of borders. This tended to be the argument and rationale put forth by contemporary intellectuals and officials, an argument that functioned to put villagers outside of history, static and unchanging. But the opposite appears to have been the case: boundary conflicts proliferated in the final decades of the nineteenth century as villagers were forced to unilaterally define village boundaries that previously had been fluid and ambiguous.[31]

This was particularly the case after 1869, when the state government adopted a draconian measure in its new land division law threatening villages with having all undivided lands declared vacant and thus the property of the government.[32] The new stipulation had little immediate impact: land divisions did not increase by any noticeable rate, and the state government would yearly issue an extension. But villagers still worried about expropriation, and they responded in different ways. Concerns over whether or not the yearly extension would be granted, and a consequent fear of potential land expropriation, led Misantla's regidor, in 1874, to recommend a quick division of the municipality's lands into four large lots according to cardinal directions.[33] In other instances, such as Chapultepec and Tonayán, local officials wrote lengthy letters attesting to why they were unable to divide their land or in which they claimed to have done so.

The prefect reviewed the case between Tonayán and Chapultepec

30 Taylor, *Landlord and Peasant in Colonial Oaxaca*, 84. My emphasis.

31 For a particularly powerful, grounded analysis of similar issues in rural, ancien-régime France, see Sahlins, *Boundaries*, chap. 4. O'Brien and Roseberry have coined the term "precipitates of capitalism" to describe processes that, while erroneously assumed to be part of "tradition," are in fact a result of a modern social transformation in the countryside. See O'Brien and Roseberry, eds., *Golden Ages, Dark Ages*, 1–18.

32 On the decree, see Ducey, "Liberal Theory and Peasant Practice," 75.

33 See Felipe Palomino to Secretario del Superior Gobierno del Estado, September 28, 1874, AGEV, RG, Tierras, Reparto, exp. de Misantla, 1872 (1874: Sección de Municipalidades, letra F, no. 13, ff. 17–19). For more on this case, see Ducey,

early in 1882. His conclusions did not bode well for Tonayán. He reported that the land division in Tonayán had not been conducted properly and village authorities had not fulfilled any of the obligations required by law. They did not submit a *padrón* (list of land recipients) to the state government for approval; they did not create a registry specifying the names of each grantee and his or her respective plot and its quality, extension, value, and perimeter; they had given land titles to some individual parcel-holders but not to others; and municipal authorities rather than state-sanctioned experts [*peritos*] had conducted the land survey. The prefect not only deemed the land division invalid but implied that in all probability it had never taken place.[34]

The prefect's remarks suggest that the authorities of Tonayán may have relied upon the *idea* and rhetoric of the land division as a means of either extending or legitimating control over marginal land or, in a modern form of the old dictum "Obedezco pero no cumplo," of perpetuating certain practices under a veneer of new dictates.[35] Villagers in Tonayán would later claim that the villagers of Chapultepec were using the division themselves as a means to lay claim to land belonging to Tonayán.[36] Such tactics were common. Villagers of Chiconquiaco accused villagers of Yecuatla of having carried out the division for the sole purpose of claiming liminal land. The *yecuatlatecos*, they argued, had even begun to rent out portions of the contested land in direct contradiction to an 1882 agreement and legal precedent.[37] Their assertion appeared to have some credence: when the yecuatlatecos contracted a surveyor (a young Victoriano Huerta) to divide their communal lands into four large lots, they excluded an article from a previous contract stipulating

"Indios liberales y liberales indigenistas," and Ducey, "Indios liberales y tradicionales."

34 Jefe Político de Xalapa to Secretaría del Gobierno, June 30, 1882, exp. Tonayán, ff. 141–43; Jefe Político de Xalapa to Secretaría del Gobierno, September 11, 1882, exp. Tonayán, ff. 145r–149v; and Secretaría del Gobierno to Gobernador del Estado, February 23, 1884, exp. Tonayán, ff. 177r–81v.

35 Similar practices seem to have been used in nearby Naolinco. See Ducey, "Indios liberales y liberales indigenistas."

36 Vecinos de Tonayán to Gobernador del Estado de Veracruz, November 10, 1904, exp. Tonayán, ff. 192–95.

37 Alcalde Municipal de Chiconquiaco to Jefe Político de Xalapa, August 9, 1886, AGEV, CLA, caja 1, exp. Tierras/Límites: Yecuatla, 1883–1920 [hereafter Yecuatla], ff. 88–91.

that land under dispute between the two communities could not be surveyed and divided.[38] In both instances, pueblos reversed the order in which the land division was to proceed, using the land division as a means to extend their holdings and assert possession and then set their borders rather than fixing their borders prior to carrying out the land division.

These conflicts were further complicated by internal village politics and client relations, and it was often certain individuals within pueblos, rather than pueblos as a unified whole, who used the border proviso to extend their holdings or their local power base. In his letter to the prefect, the mayor of Chapultepec suggested that the land division carried out by officials in Tonayán had been for the benefit of a few chosen villagers, noting that village authorities had doled out large portions of the disputed land to a mere eighteen villagers. Each received what he claimed were "immense parcels" of at least a *cuartilla* suitable for planting maize and all of which were significantly larger than other villagers' parcels.[39] The relationship of these recipients to local authorities is not clear but the mayor's insinuation provides strong presumptive evidence of how closely both the land division and border questions were related to internal community politics and clientelism. Just as suggestive is the career of Faustino Vázquez: *síndico* (the local magistrate who oversaw communal lands and land divisions) of Tonayán in 1886, Vázquez had consolidated so much land in the ensuing decades that when villagers from Tonayán sought restitution of their lands in 1917, he was the largest landowner expropriated![40]

38 See the disparity in the contracts with Antonio Guerrero and Victoriano Huerta: El Síndico del H. Ayuntamiento de este pueblo y el suscrito Antonio Guerrero, celebran el siguiente contrato para la medición y reparto de terrenos, October 15, 1883, exp. Yecuatla, f. 63r–v; and Bases pactadas entre el H. Ayuntamiento de Yecuatla y el Ingeniero Victoriano Huerta para el reparto de terrenos de dicho municipio, December 27, 1884, ibid., ff. 76–77.

39 His letter is reproduced in Jefe Político de Jalapa to Secretaría del Gobierno, September 11, 1882, exp. Tonayán. These "property owners" would write their own letter to the prefect claiming that the fixing of the border was prejudicial to their parcels. There were twenty signatures to the document. See Juan Antonio Barradas y otros vecinos de Tonayán to Jefe Político de Jalapa, November 14, 1883, exp. Tonayán, ff. 174–75.

40 Vázquez is the signatory as síndico in 1886 to an agreement between San Marcos Atexquilapam and Tonayán regarding lands. See "En la manzana de Saca-

Finally, border provisions spurred conflicts even at the regional level. The border stipulations led, as one villager observed in a letter to the governor, not only to "a struggle of pueblo against pueblo" but of "Cantón against Cantón."[41] The various prefects of Xalapa, for example, were at different times accused of favoritism toward certain municipalities within their jurisdiction and fought with other prefects, particularly those of Misantla, over cantonal borders.[42] The machinations of the prefect of Misantla led one villager to request that the government not only determine with finality the issue of municipal limits but also "definitively fix the points which indicate the dividing line between the two cantons."[43]

In other words, state fixations were not always antithetical to village interests, nor were they always unilaterally supported by the state's own agents. Indeed, at times the land division and fixed borders offered pueblos an opportunity to compel the state to support their endeavors. For example, the villagers of Chiltoyac, involved in a dispute over borders with the neighboring hacienda of Paso de San Juan, relied upon the land division as a means to win back their usurped land by notifying the governor that they could not divide their communal lands until the portion usurped by the hacienda had been returned to them.[44]

---

tal," November 17, 1886, exp. Tonayán, f. 42r–v. On Vázquez's expropriation, see Vecinos del Municipio de Tonayán to Gobernador del Estado, December 31, 1917, AGEV, CAM, Mpio: Tonayán; Poblado: Tonayán; exp. 72 [hereafter Tonayán, exp. 72]; and Ingeniero Eustolio Delgado to Ingeniero Presidente de la Comisión Local Agraria, June 12, 1931, Tonayán, exp. 72. On Vázquez's holdings see Cambrezy and Marchal, *Crónicas de un territorio fraccionado.*

41 Letter dated August 30, 1883, exp. Terrenos: Los vecinos de San José Miahuatlan. See also various documents in AGEV, RG, Tierras, Límites, 1899, exp. Límites: Lo relativo a los jurisdiccionales de los Municipios de Chiconquiaco del Cantón de Xalapa y Juchique y Yecuatla del de Misantla; and "Relativo a la propiedad que la Municipalidad de San Juan Miahuatlán dice tener en el terreno denominado Pie de la Cuesta," AMM, Caja 1882, exp. 25.

42 See for example the claims made in Juan Peña to Gobernador del Estado, August 18, 1892, exp. Tonayán, f. 64r–v; and documents in exp. Terrenos: Los vecinos de San José Miahuatlan.

43 Letter dated August 30, 1883, in exp. Terrenos: Los vecinos de San José Miahuatlan.

44 Cecilio Vázquez y Oliva y Leonardo Cortés to C. Gobernador y Comandante Militar, January 31, 1877, AGEV, RG, Tierras, Límites, 1877, exp. Lo relativo a las cuestiones que tiene sobre límites, el pueblo de Chiltoyac, Cantón de Jalapa, ff.

All these various levels of disputation and conflict served to further complicate the landscape and the state government's attempts to impose fixed borders and juridical and jurisdictional stability. As the government pushed for permanence, local inhabitants responded by appropriating the provisos and adapting them to their specific contexts. The result was ironic: the technical precision that the land division required frequently proved to be a primary impediment to its fulfillment. In order to carry out the land division, boundaries had to be firmly fixed; but attempts to firmly fix boundaries generated the very conflicts that impeded the land division. The government had seemingly tied its own hands with a self-defeating knot.

II. FUGITIVE LANDSCAPES

The judges, lawyers, and prefects charged with resolving these disputes encountered serious obstacles in attempting to end the conflicts. In the first place, most knew little about the Chiconquiacan sierra. A region of minimal intrusion by the Spanish throughout the course of the colonial era, the sierra remained socially, politically, and economically more distant than physical space would suggest. While geographically quite close to the state capital of Xalapa, the inhabitants of the sierra lived far removed from the workings of the state government. When Joaquín María Rodríguez, a well-known educator and writer from Xalapa, undertook his invaluable study of the region in 1893, he marveled at the ability of local authorities simply to fabricate data on their statistical reports to the state government with little cause for concern. Sardonically he observed that "the authorities of these remote villages know that these notices are never discussed and that no one worries about their veracity, just as no one . . . worries that it hasn't rained in Japan for another month."[45] Given the poor quality of existing roads and the ruggedness of the landscape, the sierra of Chiconquiaco may have seemed only slightly closer than Japan to officials in Xalapa. A road connecting Xalapa and Naolinco had been under construction since at least the late 1880s, but in 1893 Xalapan merchants still found it quicker and easier to

2–7. See also the case of the hacienda of San Benito and the villages of Tamalín and Tantima in Escobar Ohmstede and Gordillo, "Defensa o Despojo?" 37.

45 María Rodríguez, *Apuntes sobre el Cantón de Xalapa*, 67–68.

travel to distant Perote, accessible by railway, than to nearby Naolinco in the sierra to buy maize, and they willingly paid nearly double per *carga*.[46]

Government officials lacked textual as well as practical knowledge, which compounded the sense of distance. Governor Hernández y Hernández complained in 1871 that state officials had little more than the occasional community land title, often extremely old and imprecise, with which to determine proprietary and jurisdictional claims.[47] Others lamented the "total or partial lack of information found in the incomplete archives of [state] offices" or the "deplorable state of confusion" of the archive itself.[48] With no orthodoxy to which they could refer, judges and arbiters were compelled to make the arduous journey out to the villages themselves, hunting around in municipal halls for old documents that might aid their tasks, and often competing with village authorities trying to recover their own documents or, in one case, to gain access to a mysterious notebook described as dating from the "era of the Viceroys."[49]

Officials were often disappointed in their quests. The passage of time and the vagaries of weather both wrought their effects upon the archival record. So did the politics of pillage, a common tactic for denying a

46 On the road-building effort, see "Memoria que rinde el Jefe Político del Cantón de Jalapa al C. Gobernador del Estado de Veracruz," May 15, 1890, in García Morales and Velasco Toro, eds., *Memorias e informes*, 3:45. On the choice of Perote over Naolinco, see María Rodríguez, *Apuntes sobre el Cantón de Xalapa*, 56, who notes that a carga of maize cost eight pesos in Perote and only five in Naolinco. María Rodríguez did not mention that what constituted a carga varied considerably from place to place, although I have been unable to determine if there was significant variation in this case. With the attempt to integrate the economy and measure land, variations in weights and measures constituted a serious problem for the state, and it repeatedly attempted to impose metric uniformity, with little success. See, for example, Circular 36, Sección de Estadística, October 20, 1890; CLEV—*1890*, 155–56.

47 Address of March 9, 1886, in *Memoria leida . . . el día 13 de Octubre de 1871*, 37.

48 See Circular 9, Sección de Gobierno, CLEV—*1886*, 8–16; and J. Arizal to Manuel Acevedo, Jefe Político del Cantón [de Orizaba], November 6, 1870, AMO, caja 88, ramo de Ejidos, año 1870, exp. Noticia de terrenos desamortizados o adjudicados.

49 Juez de la Primera Instancia to Jefe Político de Misantla, December 10, 1880, exp. Terrenos: Los vecinos de San José Miahuatlan.

neighboring pueblo proof of its rights while at the same time giving one's own texts that much more value. Villagers sacked and burned their neighbors' archives or kept contested documents under tight control. They understood that historical "authenticity" and geographical legitimacy needed to be substantiated textually, a strategy not only inherited from the colonial state's obsession with documentation but also influenced by the nineteenth-century state's own attempts to centralize and "authenticate" materials in the Archivo General de la Nación.[50]

Even if found, many documents were unintelligible, faded, marred, or written in a native language. Worse still, those texts that were readable often referred to boundary markers that had long since collapsed, been cut down, or moved, or their names changed or forgotten.[51] The few maps that turned up tended to be pictographic or vernacular maps dating from the sixteenth century, such as the one shown here from Tonayán (figure 4). Although of inestimable importance and value to villagers, to state officials such maps were little more than historical curios, derisively termed *croquis* (sketch) rather than *mapa* (map).[52] They were assumed to be of dubious validity and technically worthless because they tended to represent space in sensual rather than rational terms. Documents of fundamental legal and historical import to villagers were not of interest to a state concerned primarily with technical capacity. The arbiter assigned to resolve a conflict over borders between the municipalities of Tonayán and Misantla in 1879 observed that he had only two documents upon which to base his decision: Misantla's 1791 request for a boundary survey, with certain "very old boundaries," and an early colonial map from Tonayán, which he dismissed as simply a "rough sketch with hieroglyphs."[53] He ultimately found in favor of Misantla solely on the basis of the 1791 petition. There was little uniformity in such

50 The AGN had as one of its founding principles to preserve the titles and documents pertaining to "the sacred right of property." See Palacios, "El General de la Nación, el General Agrario."

51 See for example the history of the various boundary markers in the village of Tlacolulan that, between 1823 and 1870, disappeared from the land. ACGE, exp. 7.

52 A useful discussion of the continuing use of the terms "mapa" and "croquis" in Peru to make certain kinds of distinctions can be found in Orlove, "Mapping Reeds and Reading Maps"; and Orlove, "The Ethnography of Maps."

53 Juicio de Juan Pérez, June 21, 1879, exp. Tonayán, ff. 30–31. Pérez used the word "croquis" here, which I have translated as "rough sketch."

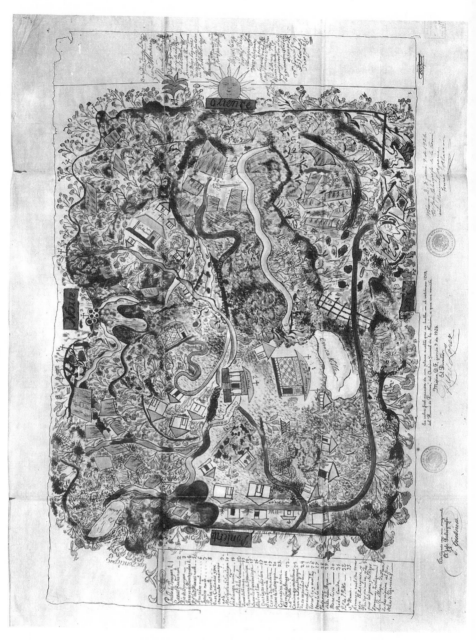

FIGURE 4. Fugitive landscapes: Anonymous map of Tonayán, no date. Courtesy of the Archivo General del Estado de Veracruz. Photograph by Bulmaro Bazaldúa Baldo.

decision making: fifteen years later the territorial pretensions of the villagers of Chiconquiaco were rejected because they had only two petitions to the viceroy, from 1590 and 1591, as supporting documentation. In light of textual dearth, the arbiter based his decision upon geographic determinants, arguing that the lands in question were closer to Juchique de Ferrer than to Chiconquiaco and that the former had better roads. In contrast, Chiconquiaco was separated from the lands by large peaks that were impassable during the rainy season, and the prefect feared that the "magnificent lands" in question would go to waste. "Before the future of the nation," he concluded, "sentiments inspired by localist pride should be quieted."[54]

A paucity of textual material impinged upon the resolution of disputes in even the most basic ways, such as determining with finality the name of a physical land feature that served as the dividing landmark between two communities. The lengthy conflict between Tonayán and Chapultepec is again a case in point. After rejecting Tonayán's claim to having divided its lands, the prefect attempted to resolve the dispute from his office in Xalapa. He initially requested that the two villages submit the appropriate maps and titles for his perusal. Although the villagers of Chapultepec submitted an unspecified set of documents, those from Tonayán requested additional time in order to send a delegation to the AGN in Mexico City to search for community documents. The prefect agreed and ordered both communities to steer clear of the disputed land. The villagers of Tonayán apparently did not heed his orders and were repeatedly warned to stop cutting forested land and planting corn in the area under dispute.[55] By September 1882 Tonayán had still not submitted any documentation, and, in November, the prefect determined that the dispute would need to go to *arbitraje* (arbitration).

Faced with outside arbitration, Tonayán and Chapultepec quickly agreed to a *juicio común*, a process in which they could mutually choose and appoint a third party to settle the matter of the border, usually a local elder or regional official thought to be well informed with regards to local history and geography. Both had compelling reasons to avoid arbi-

54 Teodoro Lecuona to Secretario del Gobierno, April 4, 1894, AGEV, RG, Tierras, Límites, 1899, exp. Límites: Lo relativo a los jurisdiccionales de los Municipios de Chiconquiaco del Cantón de Xalapa y Juchique y Yecuatla del de Misantla.

55 See the summation in the prefect's letter to the Secretaría del Gobierno, September 11, 1882, exp. Tonayán, ff. 145r–149v.

tration: in the first place, it entailed hiring nonlocal arbiters, usually lawyers from Xalapa, whose expenses were often beyond the means of most villages. In order to cover the costs of arbitration, the authorities of Chapultepec would have been forced to levy a fee of three pesos on all males over eighteen; Tonayán, meanwhile, had just recently been ordered to pay the expenses of the arbitrator in a case they had brought against Misantla (and which they had lost).[56] Both communities also realized that the process and final product of a juicio común would be at least partially under their control, malleable and negotiable, unlike the process and results of arbitraje, something confirmed by recent experience. When a conflict over borders erupted between the villagers of nearby San José Miahuatlan and Juchique de Ferrer, the mayor of Juchique firmly opposed arbitraje, stating that the villagers did not wish to expose themselves to "the trick played upon the Misantlecos who, in trusting the good faith of the arbitration judges, lost a large extent of land to the villagers of Tlacolulan."[57] Moreover, while all villagers and village authorities were allowed to accompany the arbitrator on his boundary walk during a juicio común, only the village mayors and a notary [*escribano*] could do so during a juicio arbitral.

As part of the juicio común, the two communities drew up a document that verbally described a border through references to place-names and directions: "[It is] a line beginning from the point known as 'Saltillo' situated in the Cerro Gordo, following the hillside toward 'La Gotera' and in a northwesterly direction until the point known as 'Mesa de San Pablo Guayan,' and from here taking as a departure point the eastern slope of Guayan, drawing a straight line up to the center of the hill of 'Las Chivas.'"[58] They then agreed that the mayor of nearby Naolinco, Antonio Casas, should perform the border delineation and place the boundary markers.

Casas arrived in Tonayán early on Friday morning, November 9,

56 See Municipalidad de Chapultepec to Jefatura Política de Xalapa, November 15, 1882, exp. Tonayán, f. 152r–v; and Juicio de Juan Pérez, June 21, 1879, exp. Tonayán, ff. 30–31.

57 Alcalde Municipal de Juchique de Ferrer to Jefe Político, January 16, 1880, exp. Terrenos: Los vecinos de San José Miahuatlan. Arbiters for the *juicio arbitral* were usually lawyers from Xalapa.

58 Reproduced in Secretaría del Gobierno to Gobernador del Estado, February 23, 1884, exp. Tonayán, ff. 177–81.

1883. Accompanied by a "great number of inhabitants," he walked to the Mesa of San Pablo Guayan, where he awaited the representatives and villagers of Chapultepec. With their presence, he began his work.[59] Casas did not intend his survey to be a scientific endeavor sanctified by logbooks of equations, measurements, reductions, and formulas; he was not an engineer, and there was no unpacking of measuring chains and instruments, no careful instructions to earnest assistants. Rather, Casas and the villagers would walk the land together and delineate, through reference to the *convenio*, the dividing line and place boundary markers at certain points to give the day's activities and the convenio a lasting physical permanence. Commonly referred to as an *inspección ocular*, a visual inspection of the land and a carryover from the colonial *vista de ojos*, this kind of boundary "survey" was a collective, local exercise, more description than measurement, but all the same an intensely social act of territorial definition.[60]

Casas followed the description in the convenio, and the process progressed quickly. Throughout the early afternoon he worked his way north along the eastern slope of the escarpment, following the eastern line of the hill known as the Salto de Cojolite. However, by three o'clock that afternoon a dense fog forced him to abandon his work for the day. The boundary delineation continued the following morning as the ascending sun dissipated the remnants of the previous day's mists. By eleven o'clock that morning, Casas and the villagers had placed the first boundary stone, named, undoubtedly in honor of the occasion, La Fraternidad. From this point, Casas marked out two straight lines, the first of which ran due south along the line of the Mesa de Guayan. The sec-

59 This reconstruction is based upon Casas's letter to the Jefe Político of Xalapa, November 14, 1883, exp. Tonayán, ff. 168–69.

60 In general, boundary delineation was referred to as a *deslinde;* the actual process of boundary marking was an *apeo*, which comprised both the act of marking the land as well as the resulting document attesting to the survey. See Puijol, *Guía del Propietario de Terrenos*, 7–10. Village authorities used similar terms. See, for example, the history of the boundary survey in Los Reyes, Zongolica, in ACGE, exp. 3, Cantón de Zongolica. Mariano Galván Rivera provides a lengthy elaboration on the process of the vista de ojos and measuring land in accordance with colonial instructions and legislation. In theory, the border was supposed to be measured with a *cordel*, a measuring tape of "hilo o mecate que llaman jeniquen" and of a given length, but in none of the cases examined here was any systematic measuring done. See Galván Rivera, *Ordenanzas de tierras y aguas*, 152.

ond ran north, terminating at the hill of Las Chivas, which, according to the convenio, served as a dividing landmark. And here Casas's troubles began.

When asked to identify which of the numerous hills surrounding them was that known by the name of Las Chivas, each party insisted upon the hill that extended further into the opposing pueblo's presumed territory. Unable to determine with certainty which hill was *the* hill referred to, Casas suggested that a compromise boundary be marked that would run between the two hills. Both parties agreed, and Casas concluded the boundary survey, advising each of the village representatives to come to Naolinco the following week to pick up their respective documents. However, within three days it became clear that Casas's solution would not be implemented. On November 12 he received a communication from the síndico of Chapultepec stating that in the name of the pueblo he voiced his disagreement with the delineation. Casas was surprised, and he alluded to the "hidden hand of some troublemaker [*trastornador*] of the kind that unfortunately exists in the pueblos."[61] Certain villagers of Tonayán similarly found quick reason to refute Casas's survey. Camilo Martínez, Tonayán's mayor, claimed that while Casas was "to decide which of the hills spoken of in our maps is the hill of Las Chivas," he had "no authority whatsoever" to define boundaries, construct boundary markers, and write up acts of possession, and soon thereafter a number of unnamed villagers from Tonayán destroyed the markers that had been set.[62] Martínez vociferously rejected the survey to the point of violently threatening the village síndico, Antonio Hernández, who had been the official responsible for the survey. Hernández claimed that Martínez had riled up the villagers against him by arguing that he had "lost" a portion of the entire community's land and claiming that more than sixty villagers had lost parcels in the area.[63] Regardless of hidden hands and veiled threats, the final outcome was that Antonio Casas, in a letter to the prefect of the canton of Xalapa, con-

61 Casas to Jefe Político del Cantón de Jalapa, November 14, 1883, exp. Tonayán, ff. 168–69.

62 Camilo Martínez to Jefe Político, November 19, 1883, exp. Tonayán, f. 171r–v.

63 Antonio Hernández to Jefe Político del Cantón de Jalapa, November 13, 1883, exp. Tonayán, f. 170; Camilo Martínez to Jefe Político, November 19, 1883, exp. Tonayán, f. 171r–v.

cluded that neither he nor anyone else would ultimately be able to deter-mine which was the hill "to which our ancestors gave the name Las Chivas."[64]

The case of Las Chivas was not an anomaly. Other cases between vil-lages served to emphasize the inadequacy of written boundary descrip-tions and the "interminable" problems to which they gave rise. When the arbiter in a dispute between Tonayán and Atexquilapam described and marked the boundary line, referring to "straight lines" of two hun-dred *varas* in length and "horizontal projections," not only could the pueblos not agree upon the location of the boundary marker known as La Mesita, they also argued vehemently over what constituted a vara or the straightness of an imaginary line.[65] Similar situations occurred at the level of state boundaries. In 1850 the governments of Veracruz and Puebla had reached an agreement with regard to their respective border in the sierra of Papantla. The agreement verbally described the border and referred to a "Cerro of Huipiltepec" as a primary border marker. An arbiter who had represented Veracruz in 1850 had to be rehired thirty years later, in 1879, to determine with finality which hill was "the true Huipiltepec" referenced in the agreement.[66]

Further blurring the statist vision were, ironically, a whole array of texts the government may have preferred not to have had: pueblo testi-monies and memories that rarely clarified the view. What pueblos lacked in documentation they made up for with lengthy letters of expla-nation and persuasion, deploying an idiom of engagement with the state by claiming patriotic sacrifice for the good of the nation, uninterrupted

64 Casas to Jefe Político del Cantón de Jalapa, November 14, 1883, exp. To-nayán, ff. 168–69.

65 Inspección ocular, October 11, 1887, exp. Tonayán, ff. 68–69r–v; see also the summary in Jefe Político del Cantón de Xalapa to Secretario del Gobierno, February 7, 1900, exp. Tonayán, ff. 115–19. The reference to the juicio común as "interminable" is in Secretaría de Gobierno to Jefe Político del Cantón de Xalapa, April 13, 1887, exp. Tonayán, f. 62r–v. For a similarly interesting description of a boundary inspection and the problems it faced, see the summary by Antonio Guerrero, Secretario de Misantla, in AMM, Caja 1882, exp. 25, "Relativo a la pro-piedad que la Municipalidad de San Juan Miahuatlán dice tener en el terreno de-nominado Pie de la Cuesta."

66 See *CLEV—1879*, Article 1, Decreto 55, 98–99. Similar problems plagued federal attempts to delineate terrenos baldíos. Holden, *Mexico and the Survey of Public Lands*, chap. 1, provides some illuminating examples.

peaceful possession, or both.[67] For example, authorities of the village of Tonayán tied the very absence of documents to their own national patriotic consciousness and a neighboring pueblo's traitorous use of the chaos engendered by foreign invasions. A village history of Tonayán alleged that all proof that they had purchased the lands they occupied legally in the seventeenth century had been destroyed when "Imperial forces" invaded their "humble Pueblo" in 1864.[68] The imperial forces, in this case, were not only the French but also the neighboring villagers from Tlacolulan who "took advantage of the French invasion to take our pueblo, burning the archive and taking the documents."[69] During a violent dispute over municipal limits between Yecuatla and Chiconquiaco, the mayor of the former asserted that the residents of Chiconquiaco were "mutinous usurpers, agitators, lazy and prone to drunkenness" who, unhappy with "what they have in the upper portion of the sierra, wish to descend into the fertile fields of the zona caliente." Moreover, "as everyone knows," they were disloyal, having supported Antonio Díaz Manfort's recent insurrection against the government "precisely because Manfort promised to prevent the land

67 These strategies were common throughout Mexico although the specifics often varied. In Veracruz, patriotic sacrifice by the village against foreign invaders constituted a common trope precisely because of the multiple invasions the region experienced. In the north, varying strategies of appeal tended to develop along ethnic lines: indigenous villagers usually claimed possession since time immemorial, while mestizo or white settlers emphasized their service to the nation in fighting the Apache. I am indebted to Friedrich Katz for pointing out the northern context to me (personal communication, April 15, 2000). On popular liberalism and the land division more generally, see Hernández Chávez, *Anenecuilco*, chap. 2; Mallon, *Peasant and Nation*, chap. 4; and Thomson, "'La Republique au Village' in Spain and Mexico, 1848–1888."

68 Datos que se tomaron de unos documentos muy antiguos, que a la letra dice: Estadística del Pueblo de Tonayán formado a la circular del Superior Govierno del Estado de fecha 9 de Marzo del año pasado de 1868, exp. Tonayán, ff. 395r–399v.

69 Ibid. Similar examples abound: villagers of Zongolica, in a series of documents submitted to the CGE in 1883, emphasized that they had been the victims of three different invasions by the Spanish alone and had been the first pueblo in the state of Veracruz to "second the grito de independencia." See ACGE, exp. 3, Cantón de Zongolica. Municipal authorities in Papantla reminded the federal government that "in 1862, due to the French intervention, Papantla again took up arms in defense of national integrity." See ACGE, exp. 3, Cantón de Papantla.

division."[70] In sharp contrast, the villagers of Yecuatla were "hardworking, upright, industrious and peaceful" Indians who had immediately come to the nation's defense.[71] Besides, he concluded, they had always (and still) had "legal possession from one generation to the next since before the conquest of Mexico."[72]

As a second tactic in their claims to rights, legitimacy, and priority, villagers proffered their own arguments about fixed possession by blending discourses of use rights since time immemorial with discourses of proprietary rights gained from colonial grants. They wrote extensively about their village's history and geography for state statistical collections or in lengthy, eloquent letters to the state governor. These letters tended to stress an early foundational moment by indigenous ancestors beyond the annals of history, followed thereafter by official recognition from the Spanish authorities and the granting of lands to the pueblo. A typical example is a letter by Laureano Villa, of San José Miahuatlan, written in 1879:

> It is known that in a long-ago time, whose date has been lost with the passing of the years, but which has been conserved as tradition in the memory of all the children of San José Mihuatlan because it has been transmitted from generation to generation, a number of *vecinos* of the pueblo of Chiconquiaco . . . established themselves in the fold of a hill made fertile by the waters of the río Colorado, a place located to the east of Chiconquiaco and west of Juchique, bathed by the waves of a copious river, [and] improved by the clearing of land to create fields for sowing.[73]

Officially granted portions of land by the Spanish crown "so that their *hijos* . . . could work and enjoy the land in common," the inhabi-

70 See Alcalde Municipal de Yecuatla to Jefatura Política de Misantla, May 22, 1886, exp. Yecuatla, ff. 84r–87v. Of the eighty-six known members of Díaz Manfort's rebel army, none was listed as being from Chiconquiaco. The vast majority were from the village of Colipa in Misantla. See González de la Lama, "Los papeles de Díaz Manfort."

71 Alcalde Municipal de Yecuatla to Jefatura Política de Misantla, May 22, 1886, exp. Yecuatla, ff. 84r–87v.

72 Ibid.

73 Laureano Villa y vecinos de San José Mihuatlan to C. Gobernador, December 24, 1879, exp. Terrenos: Los vecinos de San José Miahuatlan.

tants were soon forced to abandon the pueblo "to escape the effects of numerous floods, which laid waste to their fields, and of a contagious illness that decimated [*diezmaba*] them." But, Villa argued, some did not abandon the lands but rather established the municipality of San José nearby; and "we, who are their descendents, observing the same practices as our ancestors, have conserved our possession over these lands, all the rights that come with them, and it is incontrovertibly just that we continue to guard and keep those rights."[74]

Condensed into what the writer considered to be the essentials for an official audience, the history moved with little hesitation from a time immemorial—when the villagers' ancestors performed foundational acts of migration and "improvement"—to the present, when their descendants continued to work the same lands. At any given moment between those two poles, the villagers of San José Miahuatlan had purportedly enjoyed peaceful possession and, as such, had natural rights to it. The "telescoping" of time may have been in part a result of writing down an oral narrative, but it also painted a portrait of uninterrupted possession, pristine and untouched by the conflicts and disputes that otherwise riddled the countryside. In other words, the narrative's style and content was in large part conditioned by the contemporary concerns it sought to address—in this particular instance, the invasion of lands by the neighboring *juchiqueños* as well as the lack of documentation attesting to the right of possession. That there was most likely no such "golden age," characterized by peaceful and unquestioned possession, bereft of disputes or shifting use patterns, is less relevant than the fact that Villa articulated such a historical vision, one presented as received tradition, which engaged with official juridical norms regarding possession and rights, and invoked an immemorial time in place of authoritative historical texts.[75]

Not surprisingly, the villagers from Juchique countered by claiming to have also possessed the same disputed lands since time immemorial.

74 Ibid.

75 See Rappaport, *The Politics of Memory*. See also Hernández Chávez, *Anenecuilco*, chap. 4; Van Young, "Paisaje de ensueño," esp. 156–61; Ducey, "Viven sin ley ni rey"; and Nuijten, "Recuerdos de la tierra." Arguments that invoke "possession since time immemorial" had a long history in Mexico, going back to the earliest decades of Spanish settlement and conflicts over land between Indian communities and Spanish settlers. See Gibson, *The Aztecs under Spanish Rule*, 257–99.

They argued that Villa falsified the history of his village; that, in fact, the inhabitants of San José Miahuatlan were actually descendants of a different village, San *Juan* Miahuatlan. Moreover, they concluded, San José and Juchique did not even border each other: Juchique bordered Chiconquiaco on the south, not San José Miahuatlan.[76]

The historian attempting to unravel who, if anyone, told "the truth" here faces a daunting task. More to the point, so did contemporary officials. An inconsistency and confusion over names of topographical features, a lack of adequate descriptions and maps, and a morass of fragmentary and confusing land histories all conjured up a fugitive landscape—improvised, indeterminate, and administratively intangible. As it shifted in and out of focus, with it slipped away any hope officials might have had of arbitrating between disputing pueblos, defining village borders and, by extension, seeing the land division reach fruition.

III. THE FANTASY OF FIXITY

As exasperated state officials were learning, in order to implement policy, administrate, and arbitrate, the landscape itself needed to be translated into a stable, lined, and punctuated text. Fixed on the surfaces of two-dimensional maps, what appeared as practical chaos could be overwritten with discursive order. This was, in effect, both an epistemological and empirical undertaking. In the case of Las Chivas, the prefect took the initiative: faced with a dearth of reliable references (and a surplus of pliable referents) and inspired by the revolutionary promise of photography as an "absolute medium that permitted neither error nor subjectivity," he photographed the two hills in question and showed the images to various villagers, asking them to identify the hill of Las Chivas.[77] His is just one example (and a failed one at that) of an increasing

76 Alcalde Municipal de Juchique to Jefatura Política, January 16, 1880, exp. Terrenos: Los vecinos de San José Miahuatlan. See also Municipalidad de Juchique de Ferrer to Jefe Político del Cantón de Misantla, May 17, 1895, ACGE, exp. 5.

77 Jefe Político de Xalapa to Secretaría del Gobierno, December 10, 1883, exp. Tonayán, f. 165. The quoted phrase comes from a report to the Ministry of Development in "Sociedad de Ingenieros Civiles y Arquitectos: Aplicación de la fotografía al levantamiento de planos," in *Anales de la Asociación de Ingenieros y Arquitectos de México*, 95–96.

trend among state officials in the last two decades of the nineteenth century to try to endow the landscape with an official orthodoxy and textual permanence. No longer would verbal descriptions alone be acceptable; the case of Tonayán and Chapultepec adequately demonstrated the pitfalls of such forms. If state officials hoped for any success in places such as the sierra, lines of division and names of physical features had to be consistent and persist as permanently fixed signifiers, free from the vicissitudes of local memory, politics, and "customs peculiar to beings lacking a political and civil education."[78] Borders had to be not only delineated and marked on the ground, but codified on paper and archived in such a way as to create a fixed, rather than fugitive, landscape. Truly a picture could now be worth more than a thousand words.

In the penultimate decade of the century came a plethora of official decrees designed to collect fragmented and dispersed *information*, which would then be unified, structured, and rearticulated as *knowledge* to give the landscape both a physical and archival order. Prefects received instructions on how to recognize a *fundo legal* and distinguish it from other village lands subject to division.[79] Municipal authorities had to remit yearly reports to the state government listing all the places under their jurisdiction that had changed their names; they were also ordered to relinquish all "files, minutes, registries, maps, and other documents" pertaining to land division in their archives to the prefect for filing in the cantonal notary office.[80] They were given detailed instructions for the completion of a chart designed to collect information on community and municipal borders and disputes, topographical locations, placenames, and natural features such as mountains and rivers.[81]

Maps were a particular fetish in many of these orders due to the pressing need to formalize boundaries. Property surveyors were given explicit instructions to map cantonal and municipal as well as individual

78 *Memoria presentada . . . el 18 de septiembre de 1890.*

79 Circular 2, January 18, 1890, clev—*1890*, 6–8. *Fundo legal* refers to the town site, including houses and garden plots.

80 Article 40, Reglamento de la Ley General de Estadística, June 11, 1883; Article 1, Decreto 34, clev—*1889*, 210–12; reiterated in Circular 22, October 27, 1899, clev—*1899*, 52.

81 Instrucciones que deberán seguirse para llenar los esqueletos adjuntos, que servirán para correjir la carta general del Estado, October 4, 1883, amo, año 1883, caja 154, Sección de Gobierno, no. 5.

property lines and were instructed to leave copies of all their maps with the requisite prefects.[82] Moreover, the government had specific ideas about what constituted a map. As we have already seen, it could not be merely a "rough sketch with hieroglyphs"; instead, it had to reflect certain standards of accuracy, precision, and measurement. Fixed borders required fixed methods of measurement, calculated with precision. Thus by the 1880s certain adjectives begin to appear. For example, in 1886 all municipalities were required to submit to the government "a collection of topographic maps *scientifically made* in order to fix border demarcations."[83] Contracts for communal land surveys required that village limits, once described, be "fixed in a material manner" on the surfaces of maps, with directions and distances *scientifically* determined.[84] Such demands were designed to further an ideal of transparency. Meanwhile, existing laws were reiterated that severely penalized anyone destroying or moving boundary stones, trees, fences, or any other sign designed to be (or understood as) a boundary marker between properties.[85] By 1889 the state governor was optimistic: he proclaimed that while place-names could be used when necessary to locate and permanently fix boundary markers, dependence upon them would become obsolete once dividing lines had been determined and mapped.[86]

82 Contrato celebrado entre el C. General Juan Enríquez, Gobernador constitucional del Estado de Veracruz-Llave, y el ingeniero C. Manuel Gil, como representante de la *compañía* que se ha constituido para la formación del catastro parcelario de la propiedad rústica del Estado, Decreto 30, CLEV—*1891*, 122–38; Copia del contrato celebrado entre el H. Ayuntamiento de Jáltipan y el Ingeniero Albino G. Bradstreet para la medición y reparto de terrenos comunales, October 1883, AGEV, RG, Tierras, Reparto, 1903, exp. Terrenos: Lo relativo a la división y reparto de los comunales del Municipio de Jáltipan, Minatitlan. The order that all surveyors leave copies of their maps with the prefects appeared originally in Article 10 of Law 152 of May 12, 1869.

83 Circular 9, CLEV—*1886*, 8–16. My emphasis.

84 Orders given to the surveyor Ignacio Muñóz, May 2, 1902. See "En la Ciudad de Coatepec a los dos días del mes de mayo de mil novecientos," AGEV, RG, Tierras, Límites, 1902.

85 The penalty was between one month and one year in prison and a fine of between twenty and two hundred pesos. See Artículos del Código Penal, CLEV—*1881*, 398; Ley Agraria, December 14, 1881, CLEV—*1881*, 362–63; Articles 26 and 27 in Decreto 31, CLEV—*1891*, 140–49.

86 November 22, 1889, *Memoria presentada . . . el 18 de septiembre de 1890*, 79.

The governor's predictions regarding the power of the map over memory appeared to be occasionally borne out. A sharp contrast to the case of Las Chivas is the 1889 dispute over borders between Acultzingo and two communities in the sierra west of Orizaba. Even though there was not "perfect precision" with regards to the names of places referred to in the act of possession, the governor noted that state surveyors quickly resolved the problem because the dividing line had been scientifically fixed on an 1871 land division map.[87] The desire to make a claim for the efficacy of technical resolutions may have led the governor to neglect to mention, however, that President Porfirio Díaz himself had a hand in the firm resolution of the dispute.[88] Even then, conflicts were not entirely put to rest. Another dispute arose nine years later and, after consulting the same 1871 map, two engineers marked the dividing line on the ground.[89]

Although such technical solutions did not always have the immediate effect desired, they had a gradual impact on the ways in which pueblos themselves perceived boundary conflicts and their potential outcomes. Indeed, one can begin to see the development of a kind of cartographic literacy and technical language in the letters of village authorities in the last decades of the century as officials became aware of the potential a state orthodoxy might offer them in deterring conflicts or protecting their holdings. Village authorities frequently appropriated state techniques in order to better argue their case, using them as instruments of persuasion or as means to textually modernize their own authority. Recounting the 1898 resolution in Acultzingo, the municipal president eagerly reported on the spatial mastery provided by timeless coordinates, mathematically derived ground measurements, and azimuths.[90] The mayor of Yecuatla, asserting his pueblo's rights to lands claimed by Chiconquiaco, concluded a letter by reminding his official audience that he

87 *Memoria presentada . . . el 18 de septiembre de 1890*, 79. The dividing line, because it also served as part of the state boundary between Veracruz and Puebla in the mountains of Acultzingo, was originally fixed by the prefects of the canton of Orizaba, Veracruz, and the district of Tehaucan, Puebla. On the land division in Acultzingo, see chapter 3.

88 See Porfirio Díaz to General Rosendo Márquez, December 28, 1889, CPD, leg. 14, caja 27, doc. 13283.

89 "En la mojonera de la reunion," June 15, 1898, AMA, Ramo Independiente, Presidencia (1866–1949), exp. 6, leg. 1.

90 Ibid.

was "disposed to send the Government a copy of our titles and of the topographic maps that we are currently making and in no way should it be permitted that this work be interrupted because *from them and only from them* is where one can get the information sufficient for the identification of borders."[91] When the two pueblos reached an agreement in 1887, they followed the state's instructions, writing that "so that at no time will there arise doubts as to the identity of the points designated and the directions and extensions of the dividing line, added to this act is a map of the line and the adjoining lands that has been created by Señor Ingeniero don Victoriano Huerta."[92]

Regardless of such efforts, at least in the short term, most of these decrees and agreements had little effect. The majority of conflicts in the sierra of Chiconquiaco persisted beyond the turn of the century. They endured even after smaller municipalities were absorbed into neighboring municipalities by state fiat, such as when the state government "extinguished" Chapultepec and made it subject to Coacoatzintla.[93] And they outlasted a reversal of state policy: by the last decade of the century the state government increasingly began to permit communal land surveys to go forward even in places where borders had not been fixed.[94] Some indication of how little had actually been accomplished

91 Alcalde Municipal de Yecuatla to Jefe Político de Misantla, Ignacio Bentancourt, May 22, 1886, exp. Yecuatla, f. 87.

92 See the attachments to Cándido Cruz to Presidente de la Comisión Local Agraria, February 6, 1920, exp. Yecuatla, ff. 228–34.

93 The government made Chapultepec a *congregación* effective January 1, 1890. See Decreto 39, October 22, 1889, *clev—1889*, 229. Both the federal and state governments legislated certain minimum population requirements for the establishment of an *ayuntamiento* (municipal council), varying between one thousand and four thousand. This legislation, at least in Veracruz, had the political effect of reducing the number of municipios over the course of the century. On the ayuntamiento, see Ducey, "Indios Liberales y Indigenistas Liberales"; Hernández Chávez, *Anenecuilco*; and Guardino, *Peasants, Politics and the Formation of Mexico's National State*. The denomination of a place as a pueblo (and thus large enough to have an ayuntamiento) or congregación (subject to a larger pueblo) was often a product more of political relationships than of population. See Cambrezy and Marchal, *Crónicas de un territorio fraccionado*, 62–66.

94 See, for example, the instructions of the Secretaría del Gobierno del Estado de Veracruz in Alejo Galván, Jefe Político del Cantón de Coatepec, to Secretario de Gobierno, November 14, 1899, AGEV, RG, Tierras, Reparto, 1903, tomo 2, exp. Terrenos. Lo relativo al reparto de los del Municipio de Ayahualulco.

in mapping and fixing municipal limits can be garnered from the submissions of six municipalities in the canton of Misantla to the Comisión Geográfico-Exploradora in 1895. Asked to submit detailed information on and maps of the borders of its respective municipality, only one of the six (Misantla itself) sent in a map; the rest submitted traditional boundary descriptions that state initiatives had been trying for so long to overcome. For example, the authorities of Yecuatla submitted the following description: "[T]he limits of this Municipality with those of Colipa begin at a hill called 'Gueguetepec' or 'Avenencias' and from there continue north arriving at another hill called 'Yxtlahuaya' or 'del Bejuco'; from this place they run in the same direction [*por el mismo viento*] arriving at another point known by the name of 'Piedra Coyotitlán' until they arrive at the arroyo 'Yxtacapam' and here end the boundaries of Yecuatla with Colipa."[95] Such boundary descriptions were plagued by the same problems as those that had supposedly set the dividing line between Tonayán and Chapultepec and led to the dispute over the hill of Las Chivas.

The case of Las Chivas had, by century's end, achieved a kind of infamy and become the example par excellence of the importance of a fixed landscape to the practicalities of rule. In 1900 the prefect of Xalapa wrote to the governor regarding a dispute between the villagers of Tonayán and San Marcos Atexquilapam. In summarizing the case, he repeatedly referred to the continuing conflict between Tonayán and Chapultepec. He identified two primary reasons why boundaries had still not been firmly fixed nor the land division completed: "a lack of maps which detail perfectly the land" and, a partial consequence of this lack, the fact that villagers continued to assign names to physical landmarks that best served their interests:

> The identification of the name of a point that marks the boundary between two municipios, such as a hill, a riverbed, a *barranca*, etc., achieves nothing whenever there are two or more of these in the same vicinity, because when one arrives at the place under dispute, each party marks the point that serves its own interests, giving it the same

95 Ayuntamiento de Yecuatla, Cantón de Misantla to Jefe Político del Cantón de Misantla sobre límites del ayuntamiento, July 16, 1895, ACGE, exp. 5.

name as that taken as the basis for the agreement, and as this happens so often when one tries to do a boundary survey, the result is that many of these questions will never end.[96]

Nor would they anytime soon: four years later the documentation ends with a letter from a group of residents from Tonayán, arguing that "because the places in question, due to the roughness of the terrain in almost deserted and impassable mountains, are little or not known by those not born and raised in the mentioned villages, it will be difficult to find persons with the ability to resolve the question."[97] This paean to local knowledge fittingly ends the documentary record on the prerevolutionary case of Tonayán. In the end, the ambiguity over Las Chivas would only be overcome with the postrevolutionary reconstruction of the topography in the form of the ejido.[98]

CONCLUSION

Pondering the persistence of border disputes in the sierra of Chiconquiaco and the state's inability to resolve them, Xalapa's prefect argued in 1900 that they were ultimately a direct consequence of the Indian's "tenacious resistance . . . to losing even a small portion of land in order to fix the border."[99] The prefect's accusation is indicative of how difficult it was for officials to get some perspective on the literal and figurative lay of the land. Given the inadequate or, at best, ambiguous documentation available to officials attempting to fix those very borders, it is unclear to what degree they faced a "tenacious resistance" and to what degree they confused it for their own ignorance. While the evidence is admittedly

96 Jefe Político de Xalapa to Gobernador del Estado, February 7, 1900, exp. Tonayán, ff. 115r–119v.

97 Vecinos de Tonayán to Gobernador del Estado de Veracruz, November 10, 1904, exp. Tonayán.

98 See Recorrido de los linderos de Tonayán, January 7, 1932, Tonayán, exp. 72. Chapultepec would be one of the first villages in all of Veracruz to submit a petition for *restitución* in the wake of the January 1915 agrarian law. Lucio Riveros to Gobernador Comandante Militar del Estado de Veracruz, February 20, 1915, AGEV, CLA, caja 2, Tierras/Límites: Coacoatzintla, 1915–1921. They first petitioned on January 26, 1915.

99 Jefe Político de Xalapa to Gobernador del Estado, February 7, 1900, exp. Tonayán, f. 117v.

thin, it may have been that there was a very real and enduring ambiguity regarding names and locations in the sierra because agrarian practices and/or local custom had not required them to be "fixed" in the past with such specificity. Spatial formalization in the form of sharply defined borders, with a cartographic permanence, may not have been able to account for the myriad practices, uses, and relations that ultimately produced space.[100] Regardless, officials were unsettled by the very fact that, with no "maps which detail perfectly the land," they had little way of knowing. Their spatial knowledge of the countryside was contingent upon that of the rural people they purportedly governed. Throughout the latter half of the nineteenth century they struggled to flip that relationship on its head by generating a comprehensive archive of unambiguous and permanent knowledge of the landscape.

Yet the process of creating such knowledge—of marking permanent lines, plotting points, fixing place names, and making maps—proved to be hard-going. Marking, plotting, and fixing were never technical procedures distinct from a social and political context. It was for this very reason that state governors hesitated to resolve boundary disputes, at least until the mid-1890s, arbitrarily or through the use of military surveyors. Even after repeated failures in the case of Las Chivas, Veracruz's secretary of the interior suggested to the governor that "the natural thing to do would be to have the Judicial Authorities determine themselves," based upon gathered evidence and documentary proof [*constancias*], which hill should be known as Las Chivas.[101] Officials who had only recently acquired power under Porfirio Díaz could ill afford to impose their will on a whim and risk provoking rural discontent.

Just as important, people in the countryside actively participated in the procedures designed to resolve boundary conflicts. To what degree their actions constituted a "tenacious resistance," and to what degree they were simply refusals on their part to be mystified by judges, lawyers, and surveyors armed with arcane instruments, is uncertain. In either case, villagers could, and did, continuously challenge the methods through which officials sought to resolve disputes, as well as the documents they produced. Indeed, when viewed "from below," free of the

100 Lefebvre, *The Production of Space*.

101 Secretaría del Gobierno to Gobernador del Estado, February 23, 1884, exp. Tonayán, ff. 177–81r-v.

cartographer's self-delusions or the governor's illusions, maps and the processes that created them appear as ambiguous, contested, and contestable as the borders they sought to fix.

For liberal state builders committed to the division of communal lands, all of this was depressing. Regardless of the prefect's emphatic pronouncement, it was in fact difficult for regional officials to determine to what degree villagers purposefully stalled the land division, to what degree they were honest in their claims to want to divide land, or to what degree they simply found it easier and cheaper to ignore the land division rulings entirely. What does seem clear is that rural people knew enough about the state's legislation to mold their actions and words accordingly. If they wanted to, they could grind the land division process down to a halt with little more than, say, a border dispute.[102]

Any resistance that did exist was always a more complicated affair than one of communitarian villagers united against an encroaching state. The inability to permanently delineate and demarcate village borders resulted as much from an array of local conflicts and power struggles as it did from a unified community resistance to land loss. The power of caciques, for example, was contingent upon the very opacity federal officials sought to overcome. The logical consequence of fixed and mapped borders was an orthodox landscape, centrally archived, which would create a concomitant shift in the locus and reach of local power. Such power was already being challenged at times by regional prefects who had their own personal, political, and economic interests to tend to, and these often did not coincide with the state's own fixations. In other words, the standard litany of stereotypes regarding Indian tenacity and local intransigence elided the role some officials themselves played in the persistence of border conflicts and ambiguities. That same litany elided the fact that at times villagers appealed to the state to fix their borders in order to stave off the predatory encroachments of neighboring villages and large landowners or to recuperate lost land. Ultimately, the persistence of questions regarding bor-

102 Indeed, one might conclude, as has Valerie Kivelson in her work on Muscovite Russia, that at times the various efforts to fix and map the landscape "may have actually *hampered* the process of efficient state-building as much as they helped it." Kivelson, "Cartography, Autocracy and State Powerlessness," 87. My emphasis.

ders could never be easily summarized as Indian resistance nor be captured in ethnic or elitist stereotypes about either Indians or villagers. Indeed, the persistence of border ambiguities defied summary and, at least in the sierra of Chiconquiaco, the fantasy of fixity remained precisely that.

✳

CHAPTER 3

Standard Plots

On the morning of July 29, 1869, the villagers of the pueblo of Acult-zingo, Veracruz, looked from their fields with concern. There, at the edge of the Camino Nacional, which ran through the pueblo on its path between the port of Veracruz and Mexico City, stood a man armed with instruments, poring over plans and making calculations. Martin Holzinger, a Prussian engineer, had been hired by the municipal authorities to survey and divide the pueblo's communal lands into individually held plots. Absorbed in blueprints, immersed in the imaginings of a timeless geometry, his presence predicted the translation of their place into a new and, for some, foreign space. Holzinger would eventually turn what had been years of political rhetoric into physical reality. The land would be divided and community members assigned individual plots, clearly delineated, symmetrically established and numbered. The plots would be recorded on a master map archived in the municipal archive and the cantonal notary's office (figure 5).

Holzinger's map is a compelling image of the land division as well as of the mindset that promoted it. The grid of property does not sit lightly over the land but appears almost to strangle it, forcing a fractured landscape into Platonic forms. But the image is also deceiving. In presenting a smooth facade of clearly marked lines, established plots, and definitive borders, the map obscures the social process of its own production. There are no erasure marks, no smudges, no alterations. All lines are equally inscribed. This lends the map a finality that is then transposed to the very process itself, such that the land division assumes a retrospective coherence, an inevitability, it did not have in practice.

Indeed, although by sunset that first day Holzinger had completed

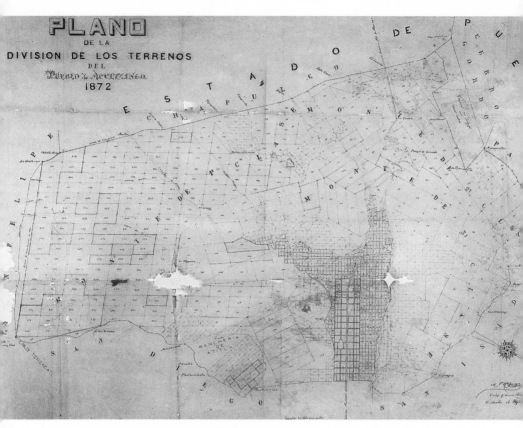

FIGURE 5. A fractured land: Martin Holzinger, *Plano de la división de los terrenos del pueblo de Acultzingo*, 1872. Courtesy of the Archivo General del Estado de Veracruz. Photograph by Bulmaro Bazaldúa Baldo.

his calculations fixing the base points for his survey, the survey (and map) took the better part of three years to complete, something Holzinger attributed in equal parts to "the lethargy of the Authorities, the apathy of the indígenas and the influence of the so-called *Tetíazcal*, who, not wanting to see the lands of this pueblo divided, devised to discredit it and put up all kinds of traps so that it would not succeed."[1] While Holzinger's survey, by his own account, constituted one of the earliest completed land divisions in the state of Veracruz, his complaints reveal just how complicated and contested the process could be. The land division was anything but the quick and easy triumph of geometry over geography.

Rarely have historians analyzed the process by which lands were divided, effectively leaving a vast void between the issuing of a governmental decree and the eventual loss of Indian land.[2] It is as if legislative text were somehow the agent of agrarian change and the future already written, a teleology turned gospel by the arrival of the revolution in 1910. But legislation was little more than wishful thinking for much of the nineteenth century. Lengthy technical interventions were required to turn the state's fixation with proprietarial transparency into a fixed reality. The land division did not take place in bureaucratic offices with state officials gazing over maps: the maps had yet to be drawn. It took place on the ground, in the *campo*, in the struggle over the very creation of those images. The surveyor, not the edict, personified the state's impending presence, and he constituted the tenuous thread running through the turbulent times that filled the space between the array of speeches and the archived map.

This chapter charts a preliminary course through this realm, following, in broad strokes, the process of land division in Veracruz in the lat-

1 M. F. Holzinger, "Informe general de la división de los terrenos comunales," May 30, 1872, AMA, libro 9 [hereafter "Informe"]. *Tetíazcal* is most likely a modified spelling in the plural of *tetiachca*, Nahua for "supreme chief of a group or tribe." See Cabrera, *Diccionario de Aztequismos*.

2 This is beginning to change. Veracruz has been the focal point for a number of recent studies of the workings of the land division process. See especially Kourí, "The Business of the Land"; Kourí, "Economía y Comunidad en Papantla"; Ducey, "Indios liberales y tradicionales"; Ducey, "Liberal Theory and Peasant Practice"; Ducey, "Tierras comunales y rebeliones en el norte de Veracruz." For detailed examinations of the process in El Salvador, see Lauria-Santiago, *An Agrarian Republic*, and Lauria-Santiago, "Land, Community and Revolt."

ter half of the nineteenth century. In part 1 I briefly survey the ideological and bureaucratic premises which underpinned the state obsession with land division. In part 2 I look more closely at how the land division was implemented and relationships between surveyors and villagers. In part 3 I examine how the logic of the land division and subsequent land distribution—the logic of a liberal "state fixation"—conflicted with local history and agrarian practice.

I. A LIBERAL VIEW OF THE LAND

"May 15, 1872," Martin Holzinger wrote, "will be a day of joy among the unfortunate *indígenas* in that pueblo [Acultzingo] who before did not even have a home."[3] On that day Holzinger completed his survey and division of the pueblo's lands, presumably putting an end to a system in which the villagers were "owners of everything . . . but possessed nothing."[4] Holzinger's assertions echoed those of John Locke who, centuries earlier, articulated the classic explanation for the presumed paradox of want in the land of plenty:

> Several Nations of the Americas are rich in Land, and poor in all the Comforts of Life; whom Nature having furnished as liberally as any other people, with the materials of Plenty, i.e. a fruitful Soil, apt to produce in abundance, what might serve for food, rayment, and delight; yet for want of improving it by labour, have not one hundreth part of the Conveniences we enjoy: And a King of a large fruitful Territory there feeds, lodges, and is clad worse than a day Labourer in England.[5]

Although more attuned to rural complexities, state officials and bureaucrats viewed communal landholdings as little better than Locke's commons. In their rhetoric, "communal" came (and has come) to obscure more than it revealed about actual land tenure and agrarian practice in the countryside. In theory, lands were owned by the municipality and worked in usufruct by villagers. Local custom and relations, power-laden and unequal, determined how it would be distributed and used.

3 Holzinger, "Informe."
4 Ibid.
5 Locke, *Two Treatises of Government*, 297.

However, land within the community was not uniformly the same, and villagers did not communally work most land. In fact, the only lands that approximated a "commons" were the ejidos—common pasture and forest near the pueblo, usually not cultivated, and initially exempt from division. That such lands were exempt suggests that one cannot make too strong a case that federal and regional officials were entirely ignorant of tenure practices in the countryside. Still, other kinds of internal complexity appear to have been either unknown or ignored by officials. There were, for example, *propios*, agricultural land set aside to be rented out for revenue; and *terrenos communales*, lands devoted exclusively to agricultural production and often divided among members of the village and farmed as family plots, with inheritance rights.[6] Such complexities were elided by officials, as were the bundle of rights, rather than notion of ownership, contained within the descriptor "communal." As a result, the term "communal" functioned not so much to describe an existing system but to describe what did not exist: individual freehold tenure and a system in which land circulated freely as a market commodity. It was, at best, a skewed representation of a complex and contextual reality that cannot simply be read through the retrospective lens of liberal theory.

Theoretically it was in the free market that the economic progress of the nation resided. Communal lands were deemed to be unproductive, unimproved, inefficiently exploited, and fiscally barren, condemned, as Alexander von Humboldt put it, to "perpetual sterility."[7] Everywhere they turned, travelers and officials saw waste, underuse, and unfulfilled potential. Journeying from the port of Veracruz to Xalapa, one traveler lamented that he had not "seen *one solitary field*, during the distance of *seventy miles*, in a state of cultivation."[8] The German traveler Carl Sartorius, on his own journey through the Veracruz lowlands offered an observation as to why: "[T]hey are not fond of hard work, nor have they any need of it, as they have plenty to live upon if they devote but a few hours a day to agricultural labor."[9] Sartorius's assumption that little la-

6 See Ducey, "Liberal Theory and Peasant Practice," and Schryer, "Peasants and the Law."

7 Humboldt, *Political Essay on the Kingdom of New Spain*, 142.

8 Siemens, *Between the Summit and the Sea*, 150.

9 Quotation from Ober, *Mexican Resources*, 17. Von Humboldt years earlier had suggested, in similar fashion, that central Veracruz could be immensely productive

bor had been necessary to ensure such bountiful yields is part of a long tradition of idealizations about agrarian labor and production.[10] Regardless, that Indians' wants were apparently satisfied by desiring little rather than producing much irritated state officials such as the governor of Veracruz, who lamented in 1870 that

> [R]ich for the conditions in which he [the Indian] is surrounded, and because he doesn't have needs, his ambition is satisfied in passing a contemplative look at a large surface of land growing flowers and varied fruits. He doesn't want nor needs more. He lives easily because at the reach of his hand he has the bread with which to eat and a piece of corn. . . . [H]e doesn't have nor want to intervene in public life.[11]

The "tragedy" of communal tenure to nineteenth-century thinkers, then, was not the twentieth-century version articulated by Garrett Hardin, who argued that a universal economic logic, rational self-interest, and demographic pressure dictated overuse (and thus "tragedy") as the inevitable result of shared property rights. Rather, to nineteenth-century liberals, the tragedy was that of underuse and inefficiency deriving from a presumed *lack* of rational self-interest and industry on the part of the Indian.[12] A lack of ordered cultivation—cleared fields, fences, property markers, even monocropped fields—corresponded to an absence of industry and simply confirmed existing suspicions: that Indians were indeed still in a veritable "state of nature." Uncultured, given to their passions rather than reason, outside of history, Indians lived in the anachronistic space of Locke's first America.[13]

---

but for "laziness—an effect of the bounty of nature" (quoted in Siemens, *Between the Summit and the Sea*, 148).

10 See Williams, *The Country and the City*.

11 Cited from Trens, *Historia de Veracruz*, 6:90–92. On fulfillment through less consumption rather than more production, see Sahlins, *Stone-Age Economics*.

12 Regardless of their diametrically opposite explanations, both arguments function to validate private property and see the commons as tragic. See Hardin, "The Tragedy of the Commons." Trenchant rebuttals can be found in Scott, *The Moral Economy of the Peasant;* Thompson, "Custom, Law and Common Right"; Thompson, "The Moral Economy Reviewed"; and Rose, *Property and Persuasion*, esp. "The Comedy of the Commons."

13 "Anachronistic space" comes from McClintock, *Imperial Leather*, chap. 1.

Liberals assumed that changing nature would change the nature of the Indian.[14] More specifically, they believed that the transformation of communally held lands into individual plots would instill a sense of acquisitiveness and promote productivity. Manuel Soto, member of the Constitutional Congress of 1857, wrote that "to give the Indians property is to tie them to the land which is theirs exclusively. Private interest will work with them to better it, and once improved it will rise in price and the desire to profit . . . will act as a stimulus to make them hard working, active and economic."[15] This was a classically liberal conception of private property that argued that only the right to exclude others would ensure that the owner alone could garner the value of his or her individual investment in the things which he or she owned. Consequently, an individual property right would encourage one to put time, labor, and care into one's land and, just as important, give the landowner a vested interest in the stability and sanctity of the state. As one state circular observed, "The governor understands that the prosperity . . . that the most chosen nations have achieved is indisputably owed to the division of property, which in forming from each small landowner a useful citizen jealous of his rights, has made productive large portions of their territories."[16] Citizens, like land, had to be cultivated.

The infatuation with land privatization cannot be reduced to a set of ideological assumptions about Indians and property, regardless of how pervasive such ideas were. There were, in fact, very pragmatic reasons to promote the land division: most importantly, communal lands were not subject to direct property taxes. The deplorable condition of the state treasury in the 1860s and 1870s inspired the government to look for new ways to increase revenue, especially property tax revenues. Communal lands in particular were targeted: once communal lands had been divided, each plot could be registered and subject to an annual direct tax, promising a substantial increase in state revenues.[17] The surveys, and

14 This is Marx's so-called dialectic of nature. See Schmidt, *The Concept of Nature in Marx*.

15 Quotation from Ducey, "Liberal Theory and Peasant Practice," 74. Such perspectives quickly assumed form as *Truth* and were repeated with mantric regularity. For an almost verbatim repetition of Soto's words a half-century later, see Carreno, "La evolución económica de la raza indígena."

16 Circular 9, Sección de Gobierno, March 9, 1886, CLEV—*1886*, 8–16.

17 See Kourí, "The Business of the Land," 233–34.

the resulting maps, would be the fundamental pillars upon which could be built a property cadaster, which in turn would facilitate taxation and property valuation at both the local and national level and "make all property perfectly marked, irreversible, and permanent."[18]

Just as important, the land division was, in James Scott's terms, a kind of "state simplification."[19] That is, it promised to distill a whole series of opaque, illegible, and contextual practices into a simple (and literal) grid of state comprehension, in the process further facilitating capitalist development. The land division thus sat at the confluence of market theory and state planning, assuring the liberal state a rationalized landscape in which property would acquire a uniform simplicity as a patchwork of bounded, abutting units. The surveyor would plot out the new plots on the land subject to division, clearing paths and erecting monuments in the process. He would then inscribe this process onto paper in the form of a map to be archived in both the municipal office and the cantonal notary. As a result, not only would state officials no longer find themselves at the mercy of conflicting claims, they could forgo fretting about nature's immanent victories over the cultural impress of the land: of collapsed mojoneras, of mangled fences, of boundary paths lost to the undergrowth. Regardless of what occurred on the ground, the system was fixed on parchment.[20]

II. PRACTICES: SURVEYORS AND VILLAGERS

The gap between rhetoric and practice could better be described as a chasm. For much of the century, the profusion of words in the numerous decrees, circulars, and laws drawn up and pronounced with the requisite magnitude—Veracruz's first land division law (1826); the federal Ley Lerdo (1856); and a multitude of others that emanated from Xalapa and Orizaba—seemed to fall on deaf ears. From 1826 to 1856 not a single

18 *Memoria presentada al Congreso de la Unión . . . 1892–1896*, 125–26.

19 Scott, *Seeing Like a State.*

20 For a particularly explicit statement to this effect, see Manuel Alvarado to Ignacio Muñoz, June 1, 1900, AGEV, RG, Tierras: Comisión Ingeniero, División de Tierras, Papantla, 1898–1905, General 2414, exp. La de ingenieros que está dividiendo en lotes los terrenos comunales del Cantón de Papantla [hereafter exp. Ingenieros].

community's land in Veracruz was divided into individual freehold parcels. In fact, there is strong evidence to suggest that in the first thirty years after independence, villagers in parts of the state actually expanded their landholdings through purchase, squatting, or outright invasion.[21] Even the years following the promulgation of the Ley Lerdo did not see a rapid increase in the number of land divisions. Words simply generated more words, so much so that in 1888 the state governor drafted legislation to consolidate, reconcile, and simplify all prior existing land division legislation.[22] Over a half-century of blueprints for a new social order appeared as little more than pompous summations of wishful thinking.

To state officials and proponents of the land division, the reason for such presumed failures was clear: Indian resistance. State governor Teodoro Dehesa succinctly summarized the views of many when he observed in 1897: "The *clase indígena* has always tenaciously opposed the division of communal lands."[23] Yet a closer look at the actual process of land division suggests that such accusations could only partially, at best, account for the lethargic pace at which land divisions proceeded. While federal and state governments were verbose about the needs for a division of communal land, and of the benefits deriving from it, they were remarkably reticent on the practical issues of how it would be implemented. Veracruz's 1826 land division law, for example, said little about very fundamental questions of who would do the surveys, pay for them, or administer the process.[24] Moreover, both the state and federal government lacked the technical, administrative, and financial ability to oversee the land divisions.[25] As a consequence, state officials eventually

21 Ducey, "Liberal Theory and Peasant Practice," 67.

22 See the reference from June 17, 1889, in *Memoria presentada . . . el 18 de septiembre de 1890.*

23 Quotation from Escobar Ohmstede and Gordillo, "Defensa o Despojo?" 26.

24 Ducey, "Liberal Theory and Peasant Practice," 74.

25 Unlike the surveying of the terrenos baldíos, which came directly under the auspices of the federal government, communal land divisions, with some exceptions, were managed by individual states and they tended to be protective over the process. On the surveying of the terrenos baldíos, see Holden, *Mexico and the Survey of Public Lands.* For considerably detailed exchanges regarding control over the pace and form of the land division in Veracruz, see Juan Enríquez, Gobernador del

determined that the practical implementation of the surveys—for example, the hiring, contracting, and paying of the surveyor—would be the responsibility of municipal authorities.[26] Such an approach had the dual advantage of not saddling the recently reinstalled liberal administration with the costs of the land division and of keeping that same administration in the good graces of the rural communities, upon whom it was dependent for support, by allowing them a substantial measure of control over the process. On the other hand, it meant that state officials put the implementation of the survey in the hands of municipal officials and into the contentious social and political world of the municipios.

From the beginning questions arose: How would the surveys be conducted? How would the surveyors be paid? What lands could be divided? Even the very basic question of who would carry out the survey proved contentious and led to repeated and lengthy delays. Land divisions, in theory, were only to be done by recognized experts [*peritos agrimensores*], a stipulation that reflected official concerns about "unscrupulous speculators" taking advantage of "the ignorance of the indígenas."[27] But such a regulation proved problematic. Despite federal stipulations regarding the examination and licensing of land surveyors (or "topographic engineers") and the existence of surveying manuals, what made one an "expert" remained vague.[28] More crucially, the absolute numbers of professionally trained land surveyors at mid-century was relatively low. Land surveying constituted one of the more popular courses of study at the Colegio de Minería in Mexico City, undoubtedly due to the relative shortness of the length of study relative to geographic or mining engineering, but the numbers were surprisingly low prior to

Estado de Veracruz, to Donanciano Lara, November 15, 1889, CPD, leg. 14, caja 25, docs. 12386–88; Juan Enríquez to Porfirio Díaz, November 15, 1889, CPD, leg. 14, caja 25, doc. 12389.

26 Decree 152, March 17, 1869.

27 Circular, State of Veracruz, February 12, 1875, AGEV, RG, Tierras, 1883, exp. Reparto de Terrenos: Lo relativo al de los de la extinguida comunidad de indígenas del Municipio de Chicontepec.

28 The federal decree regarding examination and licensing was that of January 15, 1834, and is mentioned in Galván Rivera, *Ordenanzas de tierras y aguas*, 169. Surveying manuals included Galván Rivera's text and the anonymous *Ynstrucción sobre el modo de medir tierras con expresión o explicación de los sitios de ganado mayor* (1818), BLAC, GG, ms. G371.

mid-century.[29] Even with a marked increase in the study of topographic engineering after mid-century, and particularly during the Porfiriato, when topography counted as the most popular course of study, there were still remarkably few surveyors for the entire country. The Escuela Nacional de Agricultura [ENG], created by the Ministry of Development in 1853, taught land surveying but the majority of students focused their studies on property administration.[30] Graduates who studied land surveying did not necessarily devote their efforts to *property* surveys. Indeed, it is unlikely that one made a living as a property surveyor in the nineteenth century, as many landowners viewed property surveys with ambivalence and suspicion.[31] On the one hand, as later advocates of a national property register argued, a survey and registered title would ensure the landholder complete security of property, offset potential problems with the treasury, and help guard against counterclaims and invasions with regards to property boundaries. But this may have been precisely why many landowners feared surveys and formal titling systems: a survey could mean a reduction in the size of the property held— which often had illegally expanded over the course of the years—or an increase in property valuation and taxation. Their ambivalence led the federal government to keep property registration and titling voluntary.[32]

As a consequence of such suspicions, property surveying tended to be a side project to more lucrative career paths, and officials complained of a dearth of trained surveyors.[33] Topographic engineers could find remu-

29 Mendoza Vargas, "Los ingenieros geógrafos de México," 208–9. For more details, see Craib, "State Fixations, Fugitive Landscapes," 133, table 3.1. "Geographic engineer" referred to the study of geodesy: surveying a quantity of land large enough to have to take into account the curvature of the earth's surface. Compared to topographic (or plane) surveying, geodetic surveying entailed a rigorous and lengthy course of study.

30 Bazant, "La enseñanza agrícola en México."

31 See, for example, the comments of Díaz, *Exposición internacional colombina de Chicago en 1893*, 5.

32 By 1896 only fifty-three rural properties in the entire country had been registered in the Gran Registro de la Propiedad, a clear indication of the suspicion with which surveying and formal titling was viewed by landowners. See *Memoria presentada al Congreso de la Unión . . . 1892 a 1896*, 125–26.

33 For examples, see Alejandro Prieto to Porfirio Díaz, November 9, 1889, CPD, leg. 14, caja 25, docs. 12001–2; and Mateo García y Francisco Ticante to Jefe

nerative work with the Porfirian government on the myriad development projects that proliferated, particularly after 1870, when the railroad construction boom ensued. Development projects such as these required a surveyor's technical capabilities and were more financially secure and rewarding than surveying private properties or communal lands. Holzinger is representative in this respect. His Acultzingo survey appears to have been the only communal land survey he conducted, and after its completion he quickly moved on to engineering projects in Orizaba where, as director of public works, he directed the construction of the *palacio municipal.* The city council eventually fired him in 1881, in part for a notable "lack of precision and care" in his work, but in particular for taking repeated leaves of absence to resolve "personal issues."[34] It turned out, to the council's chagrin, that Holzinger's personal issues were professional ones: he had taken a position working for the federal government as an engineer on the federal railroad in Puebla.[35]

In sum, while the state government called for divisions to be done by trained land surveyors only, this was hardly practical given their relatively low numbers. It may have been the case that, as in the nineteenth-century United States, most surveyors were self-taught, but available evidence still suggests a dearth in trained surveyors.[36] At least in Veracruz, an eclectic mix of individuals—foreigners, military mappers working for the federal Comisión Geográfico-Exploradora [cge],

Político del Cantón de Papantla, January 23, 1892, AGEV, RGJY, caja 4, general 2408, 1888–1898.

34 See Expediente relativo a la solicitud que hizo el C. Martin Holzinger para que se le aumentarse el sueldo de $45 que disfruto como Maestro de Obras de esta ciudad, January 21, 1879, AMO, OP, caja 129, año 1879; Proposición verbal de Bernardo Arzamendi, May 25, 1880, AMO, OP, caja 13, año 1880; and Licencia que se le concedió por dos meses para arreglar en México asuntos particulares, December 29, 1880, AMO, OP, caja 13, año 1880.

35 Licencia que se le concedió, December 29, 1880, AMO, OP, caja 13, año 1880; M. F. Holzinger, Ferrocarril de Puebla a Izúcar de Matamoros, Sección 2a. de Cholula a Atlixco—20. tramo, July 15, 1882, in *Memoria presentada al Congreso de la Unión . . . enero de 1883 a junio de 1885*, tomo 6, *Atlas*.

36 See, for example, the remarks of the prefect of the Cantón of Papantla in Mateo Garcia y Francisco Ticante to Jefe Político del Cantón de Papantla, January 23, 1892, AGEV, RGYJ, caja 4, general 2408, 1888–1898. On U.S. surveyors being self-taught, see Faragher, *Sugar Creek*, 39–40.

and municipal officials—surveyed communal lands.[37] Some, such as Holzinger and surveyors from the CGE, were professionally trained engineers and knew at least the basics of surveying; others did not, particularly those village officials and regional elites chosen more for questions of money, personal interest, or convenience than for their abilities. As a result, land division surveys were often carried out by individuals of questionable training, skill, or impartiality, and with little consistency, a supposition borne out by the existing land division and property maps archived in the Mapoteca Orozco y Berra.

In the 1890s the CGE began collecting and archiving local property and land division maps of the regions in which it was working. The office personnel filled the maps with marginal comments lamenting the poor quality and minimal value of the images. To mention but one example, Bernardo Mayer's 1883 map of Misantla, archived in 1892, could only be used "for details on the names of hills," as the plotted positions were entirely at odds with the most recent traverse surveys done by CGE engineers.[38] The comments of successive surveyors on the work of their

37 A compatriot of Holzinger's, Bernardo Mayer, surveyed the communal lands of Misantla. There were also a number of Italian and U.S. citizens surveying communal lands throughout Veracruz. Meredith Jones mapped Misantla's municipal boundaries. On Mayer's work in Misantla, see the summation from July 10, 1924, AGEV, CLA, caja 4 (1922–1929), exp. Misantla, exp. Comisión Local Agraria: Asunto, ff. 472–74. Mayer mentions he is from Prussia in Bernardo Mayer to C. Gobernador del Estado, May 20, 1888, AGEV, RG, Tierras, Reparto, 1872, exp. de Misantla, ff. 193–94. Mayer's map of the canton can be found in MOB, CGV, Varilla 7, no. 6423. On Jones, see H. Ayuntamiento de Misantla to Jefe Político del Cantón de Misantla sobre límites del Ayuntamiento, June 27, 1895, ACGE, exp. 5. See also documents relating to Alvin Bradstreet and F. O. Harriman in AGEV, RG, Tierras, Reparto, 1903, exp. Terrenos: Lo relativo a la división y reparto de los comunales del Municipio de Jáltipan, Minatitlán. Once the CGE relocated its base of operations from Puebla to Xalapa in 1881, CGE explorers and surveyors are found in increasing numbers contracting their services to communities. For more on the CGE, see chapters 4 and 5.

38 Cantón de Misantla, por Mayer, December 8, 1883, MOB, CGV, Varilla 7, no. 6423. See also Mapa del Mpio. de Misantla, Ing. Meredith Jones, MOB, CGV, Varilla 7, no. 6420. Private maps of haciendas were also unreliable. See for example, Plano de la H. de 'Cofradia' que representa las divisiones del terreno que practicó el Sr. M. Nuñes, MOB, CGV, Varilla 4, no. 6200; Plano de la Hacienda del Jazmín, MOB, CGV, Varilla 6, no. 6331; Plano de la Hda. de Sn. Cristobal, levantado por el Ing.o.

predecessors are similarly revealing. Ignacio Muñoz, Porfirio Díaz's nephew and head of the land division in Papantla in 1896, complained that the map resulting from a survey twenty years prior was "very inexact" and "completely erroneous and arbitrary."[39]

Regardless of the paucity of qualified surveyors, village officials, under the threat of expropriation stipulated in a state decree of 1869, eventually contracted with someone to survey village lands.[40] By no means was there consensus within pueblos—rife with generation, gender, and class conflict—regarding the selected individual. Indeed, the process of selection often proved to be lengthy and contentious, as disputes arose between different factions within pueblos and even between municipal authorities themselves.[41] A particularly detailed example is provided in the records from Ixhuacan de los Reyes, a municipio in the canton of Coatepec, where a dispute over the surveyor lasted some three decades.[42] When the municipal president, over the opposition of the síndico who oversaw the process of land division, appointed a local community member to carry out the survey, arguments arose. Soon after, when the newly appointed surveyor attempted to begin his work, a number of Indians in the town took to the streets shouting (somewhat paradoxically) "Death to the authorities, long live Sr. Síndico."[43] Al-

---

militar José Ygnacio Ibera en 15 de julio de 1834 a la escala de 1:60,000, MOB, CGV, Varilla 9, no. 6569.

39 Ignacio Muñoz to Gobernador del Estado, March 5, 1896, exp. Ingenieros. For other examples, see Ducey, "Liberal Theory and Peasant Practice," 67.

40 Decree 152, March 17, 1869.

41 Splits within municipal councils are not surprising: the power of indigenous leaders in communities did not simply vanish by official fiat with the abolition of the Republic of Indians, and they often sat on the municipal council. The German traveler Carl Sartorius observed the following while traveling in Veracruz: "The Indians are all citizens of the republic, and choose their municipal authorities according to law. In the genuine Indian villages, nevertheless, one cannot fail to observe the existence of the authority of certain aristocratic families, whose credit is rendered sacred by custom, and whose decree in all local matters is considered decisive. They keep the flocks together, manage the funds of the community (often to their own advantage), influence the choice of the communal authorities, punish the youth, and arrange marriages." Sartorius, *Mexico*, 67.

42 AGEV, RG, Tierras, Reparto, 1885–1911, exp. Ixhuacan: Lo relativo al reparto de los comunales del municipio de Ixhuacan [hereafter "exp. Ixhuacan"].

43 Alcalde Municipal de Ixhuacan to Jefe Político de Coatepec, December 31, 1885, exp. Ixhuacan.

though quelled quickly, the brief riot prompted a visit by the prefect, who eventually brokered an agreement in which the surveyor would remain but his work would be subject to personal review by the síndico.[44] The prefect's reasoning in this instance is especially interesting: the protestors were not angered by the land division per se but rather by the imposition of a surveyor whom they claimed was not impartial. His solution forestalled any further immediate violence, but by the turn of the century the town's lands had still not been divided, and in 1905 military surveyors from the CGE were ordered in to do the task.

Questions regarding the surveyor often transcended village limits. Regional elites and state officials had their own personal, political, and economic interests at stake and wielded influence over the assignment of the surveyor. A military surveyor who became prefect in the canton of Misantla, for example, assumed complete control of the process, appointing a former colleague to survey the undivided lands under his jurisdiction. His colleague was well paid from the village coffers, while the prefect ended up with a number of plots of the land.[45] In Acultzingo, when the municipal authorities were delinquent in paying Holzinger, an unspecified merchants association [*Comercio*] offered him a weekly fee of twenty-five pesos, provided he also divide portions of *monte* (wooded) land of interest to them not included in the original division. Holzinger agreed, claiming it to be "the only means of completing the reparto, because if I waited for the Authorities . . . it might never have been completed."[46]

Regional officials and elites disputed the surveys among themselves. For example, prefects on occasion would reject a contract because the surveyor was associated with a political enemy or faction. The prefect of Misantla rejected a village's choice of surveyor because the surveyor's *fiador* (guarantor) was the previous prefect, with whom he had a personal conflict. He thus imposed a new surveyor at substantially increased cost to the village.[47] At times these complicated alliances and animosities led

44 Jefe Político Jacinto García to Secretaría del Gobierno, April 6, 1887, exp. Ixhuacan.

45 See Francisco Canóvas to Secretaría del Gobierno, July 7, 1902, AGEV, CLA, caja 1, exp. Yecuatla, ff. 172–74r–v. Canóvas was the recipient of six lotes in the division. See ibid., ff. 223–24r–v.

46 Holzinger, "Informe."

47 See AGEV, CLA, caja 1, exp. Yecuatla, ff. 62–77, esp. 68.

to violent confrontations. One brutal instance comes from southern Veracruz in 1887. There a merchant, Anastacio Elejalde, wrote Porfirio Díaz complaining that the prefect had forced an indigenous community to contract with an Italian surveyor at an inflated price of $20,000 which, along with forced Indian labor, allowed the surveyor to live in luxurious comfort in a "feudal encampment" decorated in Italian flags.[48] Elejalde, along with his wife, young son, and six servants, wound up gruesomely murdered as did his associate, Francisco Hernández, shot dead in the "shadows of the night."[49] The surveyor, Victor Assennato, and his associates were the primary suspects.

Why such concern over the surveyor? After all, were they not, with their instruments and logbooks, merely mechanically implementing the state's plans? The contentious process that surrounded the selection of the surveyor calls such rationalist notions into question. Measuring land in the countryside was not like measuring quantities in the laboratory.[50] The practice of the survey unfolded in time, in the heat of the fields under the eyes of the villagers. Surveyors, regardless of qualifica-

48 Anastacio Elejalde to Porfirio Díaz, January 13, 1886, CPD, leg. 11, caja 1, docs. 409–10.

49 *Apuntes privados de las causas que ocasionaron los asesinatos de San Pedro Soteapam del Cantón de Acayucán, Estado de Veracruz, el 10 de Mayo de 1888*, anonymous letter to Porfirio Díaz, CPD, leg. 13, caja 15, docs. 7380–81. Elejalde's relationship with Assennato was more complicated than he let on. In an anonymous letter to Díaz, Elejalde is described as an intimate friend of Assennato's whose relationship soured. While Assennato was presumably never indicted or tried for his supposed crimes, he did leave the area soon after, moving north to perform surveys in Papantla and Ixhuacan in 1899, at which point he disappears from the archival record. See exp. Ixhuacan; MOB, CGV, Varilla 10, no. 6609; and Teodoro G. y Lecuona, "Memoria que rinde el Jefe Político del Cantón de Xalapa al C. Gobernador del Estado de Veracruz," May 28, 1895, in García Morales and Velasco Toro, eds., *Memorias e informes de jefes políticos*, 3:97.

50 On the ways in which science and laboratory practices legitimate power-laden activities as somehow outside the realm of politics, see the collected essays in Lenoir, ed., *Inscribing Science;* Latour, *Science in Action;* and Latour and Woolgar, *Laboratory Life*. For a contemporary and devastating critique of the development industry and its self-promotion as a neutral, scientific agency, see Ferguson, *The Anti-Politics Machine*. I leave aside for now a whole series of questions regarding actual acts of measurement, which gave rise to a host of problems. See the classic work of Kula, *Measures and Men;* and Thompson, "The Moral Economy of the English Crowd in the Eighteenth Century."

tions, were subject to the influences, threats, and overtures of those around them, whether they be the opportunistic offerings of the merchants of Acultzingo or the anonymous note posted on a surveyor's door in Misantla: "Surveyor, my friend, go back to where you come from because if you stay, we'll meet you in the monte one day."[51]

Moreover, surveyors were not passive extensions of objective instruments. They came to the field with their own politics, persuasions, and interests. For example, the antagonism directed at Victor Assennato by large landowners outside Acayucán may in part be explained by his own admission that he worked "only to contribute to the well-being of my adopted country and for those known as my brothers, the *hijos* of this pueblo [Acayucán]."[52] Although Assennato's claim should not be taken at face value, the fact remains that when he "translated" the old and imprecise land measurements contained in the villagers' colonial titles into contemporary dimensions so that they might set the boundaries of their community and recuperate lost land, he sparked the ire of a powerful, neighboring large landowner, who claimed that his modern interpretation of the village's colonial landholdings was inaccurate and the result of payoffs.[53] In another instance a surveyor was imprisoned as an enemy of the state for purportedly inciting the Indians in Papantla to rebel.[54] The eclectic and diverse array of individuals who were hired to perform surveys in Porfirian Veracruz prevents any easy assignation of ideological coherence. As the case of Assennato suggests, the surveys could at times prove as undesirable to *hacendados* as they are commonly construed to have been for villagers. Moreover, what is clear is that surveying could not be carried out in a controlled environment. While a surveyor's primary task may have been to observe, measure, and describe

51 Quotation from Ducey, "Indios liberales y tradicionales," 16.

52 Victor Assennato, Relación de la exploración de los terrenos municipales de este muncipio, March 21, 1887, AGEV, CLA, caja 1, Acayucán: 1903–1921, exp. Restitución de ocho sitios de ganado mayor que reclaman los naturales de las Congregaciones de Acayucán [hereafter exp. Acayucán], ff. 144–46.

53 Oficio 4, 641, de la Secretaría de Fomento del Gobierno de Veracruz, April 25, 1887, in exp. Acayucán, f. 146.

54 See the case of Severiano Galicia, discussed in Chenaut, "Fin de siglo en la costa totonaca." Galicia was accused of similar activities—of provoking the Indians to "a revolution of a socialist character"—during his surveys in Amecameca, Puebla, in 1890. See Aboites and Morales Cosme, "Amecameca, 1922," 70–78.

physical reality, a series of tasks that create an impression of removed study, rarely could the surveyor feel or be so aloof. Surveyors did not observe a static and natural landscape: they both worked in and recreated a social, cultural, and political world.

Once the surveyor had been chosen, municipal officials signed a contract with him. While they varied from place to place, certain clauses appeared with regularity in the contracts examined.[55] In the first place, village authorities tended to assume responsibility for providing the

55 Based upon Holzinger's notes in his "Informe" and the following surveying contracts: in AGEV, CLA, caja 1, exp. Yecuatla: El H. Ayuntamiento de este pueblo [Yecuatla] y el suscrito Antonio Guerrero, celebran el siguiente contrato para la medición y reparto, October 15, 1883, f. 63r–v; Bases pactadas entre el H. Ayuntamiento de Yecuatla y el Ingeniero Victoriano Huerta para el reparto de terrenos de dicho municipio, December 27, 1884, ff. 76–77; Bases para el contrato que debe celebrarse entre el H. Ayuntamiento de Yecuatla y el Topógrafo D. Agustín Carranza para la división de los terrenos que formen el ejido de dicho pueblo, January 26, 1895, ff. 105–7; Contrato celebrado entre los Señores Juan C. García y Ingeniero José B. Barroeta para el levantamiento y división de los terrenos que forman el Ejido de este pueblo, June 3, 1895, ff. 112–13r–v; Untitled contract with Ignacio Muñoz dated June 27, 1902, ff. 173–74r–v.

In AGEV, RG, Tierras, Reparto, 1872, exp. de Misantla: El H. Ayuntamiento de Misantla celebra con los C. C. Ingeniero topógrafo Bernardo Mayer, Felipe Palomino y Patricio Zorrilla . . . , March 20, 1882 (letra R [1879–1884], Reparto de Terrenos: Lo relativo a las disposiciones dictadas para el de los que en común poseen los vecinos del Municipio de Misantla, f. 81r–v); Contract between Colipa and Ramón Córdoba, July 14, 1883, exp. Terrenos. Lo relativo al reparto de los comunales del Municipio de Colipa, Misantla, ff. 3–4; El C. Síndico del H. Ayuntamiento de este pueblo [Colipa] . . . con el Antonio Guerrero el siguiente contrato . . . , September 20, 1883, exp. Terrenos. Lo relativo al reparto de los comunales del Municipio de Colipa, Misantla, f. 13r–v; Bases pactadas entre el H. Ayuntamiento de Colipa, el Sr. Ing. D. Victoriano Huerta, el Sr. D. Manuel Viveros y el Sr. D. Francisco de la Hoz para llevar a efecto la mensura y reparto de los terrenos de comunidad pertenecientes al mencionado pueblo, June 20, 1884, exp. Terrenos. Lo relativo al reparto de los comunales del Municipio de Colipa, Misantla, ff. 22–24; Contrato celebrado entre el Ayuntamiento de Colipa y el Ingeniero Gil Manuel [Manuel Gil], para el fraccionamiento del Ejido del pueblo de Colipa, November 9, 1896 (no. 73, letra E, Ejido. Lo relativo a la división y adjudicación del de Colipa del Cantón de Misantla, f. 2); Contrato celebrado entre Juan Manuel Hernández y el Síndico de este Ayuntamiento [Colipa], February 4, 1907 (no. 6, letra F, Fundo Legal. Lo relativo al del Municipio de Colipa, f. 2r–v).

In AGEV, RG, Tierras, Reparto, 1903: Copia del contrato celebrado entre el H. Ayuntamiento de Jáltipan y el Ingeniero Albino G. Bradstreet para la medición y

surveyor with men to cut down brush and open the lines of sight to facilitate measurement. If the surveyor was not a member of the community, municipalities also covered the surveyor's expenses during the survey and usually housed and fed him. There appear to have been few complaints by villagers at having to clear brush and carry the surveyor's instruments. However, the housing and feeding of the surveyor could, and did, create problems. For example, in 1892 villagers from Papantla wrote the prefect claiming that while they wished to proceed with the land division, they did not want it to be done by the surveyors from the canton, and particularly by one Salvador Martínez who reportedly had required other villagers to "maintain [him] in a luxurious manner."[56]

Peccadilloes such as these took on greater significance in places such as Papantla when those very same surveyors shirked their duties. In general, the surveyor's responsibilities included performing the survey within a certain time frame, usually varying from two to six months. This meant not only measuring the land into plots according to the number of stated recipients but also creating a final map of the survey—with information on the extension of land measured and the number and size of plots clearly demarcated—for deposit in the municipal archives.[57] This latter stipulation was critical. For one thing, the map effectively represented the new division of land for use in the municipal registry and for the assignment of plots and the resolution of disputes. Moreover, village authorities wanted to ensure that after the land division they would not be dependent upon sanctioned experts to make

reparto de terrenos comunales, October 1883, exp. Terrenos. Lo relativo a la división y reparto de los comunales del Municipio de Jáltipan, Minatitlan.

In AGEV, RG, Tierras: Comisión Ingeniero, División de Terrenos, caja 1, general 2414, 1895–1905, exp. La de Ingenieros que está dividiendo en lotes los terrenos comunales de Papantla: Contrato entre la Junta Directiva del Lote denominado antiguamente San Martín, hoy Troncones y Potrerillo, sito en el Municipio de Coazintla y el Ingeniero Don Herculano Martínez, n.d.

In FCP: Untitled contract between Francisco Canóvas and the Ayuntamiento of Cosamaloapam, May 19, 1914.

56 Mateo García y Francisco Ticante to Jefe Político del Cantón de Papantla, January 23, 1892, AGEV, RGYJ, caja 4, general 2408, 1888–1898. Interestingly, García and Ticante mentioned Victor Assennato as a trustworthy and potential surveyor.

57 In certain instances, he would also be responsible for determining the value of land, but this was not consistent.

sense of new proprietary configurations. They recognized that the constitution of a new set of social "facts" had the potential to dramatically curtail their power. Surveyors were thus required to mark the land division plots "distinctly and in a convenient scale on the map, leaving on the land points of natural or artificial reference easily recognizable and identifiable" so that the material demarcation of each land plot could be known "without the presence of the Engineer and by a person without scientific knowledge."[58]

Just as important, the final map proved that the village had indeed legally divided its lands and testified as much to the state government.[59] Without the final map, villagers were condemned to resurveying their lands at their own expense. The case of Victoriano Huerta constitutes one of the more notorious in this regard. Huerta, a military officer who would eventually go on to overthrow Francisco Madero in the early stages of the Mexican revolution, contracted his surveying services to a number of villages in central and northern Veracruz in the 1880s. Charged with dividing their communal lands, Huerta repeatedly failed to fulfill his obligations by leaving the villages upon receipt of payment without completing the survey or producing a final map of the division. Papantla's prefect in 1895 reported that a primary reason for increased opposition to the land division there was Huerta's previous deceit: co-owners of large communal plots had laid out the substantial sum of $5761.25 with no results.[60] Ironically, Huerta had been sent to Papantla in part because another surveyor—the aforementioned Salvador Martínez—had himself been accused of not completing his duties and of trying to charge land recipients for the maps of their parcels.[61] Repeated in-

58 See Bases pactadas entre el H. Ayuntamiento de Colipa, el Sr. Ing. D. Victoriano Huerta, el Sr. D. Manuel Viveros y el Sr. D. Francisco de la Hoz para llevar a efecto la mensura y reparto de los terrenos de comunidad pertenecientes al mencionado pueblo, June 20, 1884, AGEV, RG, Tierras, Reparto, 1872, Exp. de Misantla: exp. Terrenos: Lo relativo al reparto de los comunales del municipio de Colipa, Misantla, ff. 22–24.

59 According to Article 10 of Law 152, May 12, 1869, a copy of the final map was to be sent to the prefect so that it could be archived in the cantonal notary. This proved also to be the case for the surveys of terrenos baldíos. See Manero, *Documentos interesantes sobre colonización*, 63.

60 Angel Lúcido Cambas to Secretario de Gobierno, October 26, 1895, exp. Ingenieros.

61 Mateo García y Francisco Ticante to Jefe Político del Cantón de Papantla,

stances such as these increased villagers' suspicions about the entire land division process, particularly when their repeated complaints to state officials garnered only a perfunctory and unsatisfactory response.[62]

Survey contracts seem to capture these suspicions. While early contracts generally gave the surveyor a flat fee for his work, later contracts adopted a system whereby the surveyor would be paid a third at the outset, a third at midpoint (usually upon the completion of the actual process of measurement), and the final third only upon submission to the municipal authorities of the final map or a system in which the surveyor would only be paid upon completion of the survey and the approval of the map by the government.[63] While the evidence is limited, such changes strongly suggest that municipal authorities learned, often through bitter experience, to draw up the contract in such a way as to protect themselves, particularly from surveyors who received payment upon completing their fieldwork but neglected to leave a copy of the finalized map of the division.

Contracts also reveal an increasingly "modern" cartographic literacy among municipal authorities, who began to make demands about the appearance of the maps. For example, in 1883 the síndico of the village of Colipa signed a contract with Antonio Guerrero, in which the only mention of a map was "a sketch that gives an idea of the extent of the land measured."[64] In contrast, in a 1907 contract with Juan Manuel Hernández, the síndico not only specified that a general map of the division

January 23, 1892, AGEV, RGYJ, caja 4, general 2408, 1888–1898; Angel Lúcido Cambas to Secretario de Gobierno, January 23, 1896, AGEV, RGYJ, caja 4, general 2407, 1894.

62 See for example Alcalde Municipal to Jefe Político, August 15, 1896, AGEV, CLA, caja 1, exp. Yecuatla, ff. 129–30; and T. M. Paredes to A. Lucido Cambas, September 14, 1892, AGEV, RGYJ, caja 4, general 2408, 1888–1898.

63 Compare, for example, the general language and timing of payment in earlier contracts, such as Holzinger's and Guerrero's, with that of later contracts. See the contracts listed in note 55. For an example of the tripartite payment schedule, see Bases pactadas entre el H. Ayuntamiento de Yecuatla y el Ingeniero Victoriano Huerta, December 27, 1884 exp. Yecuatla. Surveyors hired to divide Yecuatla's ejido land in 1895 and again in 1902 were paid by the hectare but only after submitting the final map of their work and having it approved by the government. See the contracts with Carranza, January 26, 1895, exp. Yecuatla, ff. 105–7 and Muñoz, June 27, 1902, exp. Yecuatla, ff. 173–74r–v.

64 El C. Síndico del H. Ayuntamiento de este pueblo [Colipa] . . . con el Anto-

as well as individual maps of each plot be created; he also insisted that the maps be "elegant and in color."[65] In other instances, authorities required the surveyor to clearly mark on the map that he had adjusted his measurements to account for the declination between true and magnetic north. Municipal officials were gradually learning the rhetorical codes of persuasion that saturated the mapped surface, codes that ensured the map's status as bonafide map [*mapa*] rather than dubious sketch [*croquis*].[66]

A critical question municipal officials had to answer at the time of signing the contract was how the surveyor would be paid. They had a number of options: levy a fee on all villagers who would be receiving plots; sell portions of the village's land; give the surveyor portions of the land surveyed; or use a combination of these options.[67] Regardless of

---

nio Guerrero el siguiente contrato, September 20, 1883, AGEV, RG, Tierras, Reparto, 1872, exp. de Misantla, exp. Terrenos: Lo relativo al reparto de los comunales del municipio de Colipa, Misantla, f. 13r–v.

65 Contrato celebrado entre Juan Manuel Hernández y el Síndico de este Ayuntamiento [Colipa], February 4, 1907, AGEV, RG, Tierras, Reparto, 1872, exp. de Misantla (no. 6, letra F, fundo legal: lo relativo al del Municipio de Colipa, f. 2r–v).

66 For one example of how a map could be rejected as documentary evidence by disparaging it as a croquis, see chapter 2. A useful discussion of the continuing use of the terms "mapa" and "croquis" to make certain kinds of distinctions can be found in Orlove, "Mapping Reeds and Reading Maps," and Orlove, "The Ethnography of Maps."

67 Holzinger, for example, received a flat fee for his work and a *solar* (garden plot) in the fundo legal at the western end of the Camino Nacional; in another case, the community offered the surveyor land at a reduced price per hectare in addition to a base fee for the survey. See Holzinger, "Informe"; Holzinger, "Informe de la División [*sic*] de Terrenos: Demostración de los gastos erogados a la fha. y presupuesto para la conclución [*sic*]. Proyecto de Divición [*sic*] del Pueblo en Solares, Acultzingo, 1871," AMA, Presidencia, libro 4; Copia del contrato celebrado entre el H. Ayuntamiento de Jáltipan y el Ingeniero Albino G. Bradstreet, October 1883. AGEV, RG, Tierras, Reparto, 1903, exp. Terrenos: Lo relativo a la división y reparto de los comunales del municipio de Jáltipan, Minatitlan. Bernardo Mayer carried out the land division of the municipal lands of Misantla between 1882 and 1884, supposedly in return for a sitio de ganado mayor. Originally he had been contracted by the pueblo in 1875 at the recommendation of the prefect, but the ayuntamiento claimed not to have the money to pay him. Alcalde Municipal de Misantla to Jefe Político de Misantla, February 23, 1875, AGEV, RG, Tierras, Reparto, 1872, exp. de Misantla, f. 68r–v and "En la villa de Misantla" in ibid. The

these choices, forcing pueblos to cover the costs ensured surveys would proceed only haphazardly. The costs constituted an enormous burden, particularly for smaller villages that complained that they did not have the funds available to pay the engineer. In addition, municipal authorities, while not necessarily opposed to the land division, did not assign it the same priority as state officials. In Acultzingo, Holzinger accused the municipal authorities of being "lethargic" with respect to the land division, and he eventually turned to a group of unnamed merchants for financial recompense. Municipal authorities often opted to use available funds to construct municipal buildings or purchase portions of land rather than pay for a surveyor. Michael Ducey has shown, for example, that while communities claimed not to have the money to carry out the land division, they could come up with it quite quickly for the purchase of land.[68]

Once a survey began, payment remained a source of constant interruption and delay. Holzinger reported that "I had to *suspend the work because the Municipality failed to comply with Article 4 of the contract*, which stated they were to give me partial payment for my expenses."[69] As we have already seen, he eventually received payment from a group of merchants instead. Even then, the year-long postponement meant that, due to the prolific growth of the flora during the rainy season, he had to reclear all his lines of sight and reestablish his reference points, delaying the survey further.[70]

In other cases the issue of payment sparked a whole series of questions within and between pueblos about exactly whose land could be divided and under whose jurisdiction. The municipal authorities of Ayahualulco (canton of Coatepec), considering the "immense poverty in which everyone found themselves," voted to sell a portion of land to pay for the land division rather than levy a fee on each villager.[71] The land they chose to sell was, conveniently, located in the village of Setlalpam, a *su-*

---

alienation of a parcel of land to pay for the survey was permitted by Article 13 of the Ley de Subdivisión de Propiedad.

68 Ducey, "Indios liberales y liberales indigenistas."

69 Holzinger, "Informe." Emphasis in the original.

70 Ibid.

71 "En el pueblo de Ayahualulco del Cantón de Coatepec a los diez y seis días del mes de Noviembre," in Copia certificada de las constancias que existen en la oficina de esta Alcaldía, relativa al reparto de los terrenos de comunidad de este

jeto within their political jurisdiction. The villagers of Setlalpam immediately appealed to the state governor, claiming the land belonged exclusively to them.[72] They argued that they should be able to survey and divide their lands independently from the municipal seat of Ayahualulco:

> [A]ccording to tradition, for more than sixty years the vecinos of this Congregación have possessed exclusively these lands without the intervention of the vecinos of Ayahualulco; and it could not be any other way, because nature itself has made this division obvious; effectively, the lands of Setlalpam are completely separated from those of Ayahualulco, such that between this Congregación and Ayahualulco sits the Municipio of Ixhuacan in such a way that between the two places there is a distance of five leagues.

"In view of the topographic situation of these places," they thus asked the government to resolve the dispute in their favor.[73] The head of Veracruz's Ministry of Development concurred, stating that while Ayahualulco had jurisdictional authority over the lands, it did not have rights to actual property: "[T]he question of jurisdiction," he wrote, "is in no way connected to that of property."[74] Setlalpam eventually won the right to have their lands divided separately.

The case of Setlalpam and Ayahualulco is exemplary of the utter complexity and opacity the government confronted on the ground when attempting to bring the land division to fruition. Its dependence upon local structures of power to implement the land division meant that it inherited all the local conflicts, rivalries, and histories that inevitably foiled the creation of a simple grid of uniformity. By putting the implementation of the process into the hands of the municipios, state officials may have made a financially savvy choice and saved themselves the po-

---

Municipio, AGEV, RG, Tierras, Reparto, 1903, exp. Terrenos: Lo relativo al reparto de los del Municipio de Ayahualulco [hereafter exp. Ayahualulco].

72 Originarios de la Congregación de Setlalpam to Co. Gobernador, September 20, 1893, exp. Ayahualulco; Acuerdo, February 23, 1893, exp. Ayahualulco; Jefe de la Sección to Gobernador del Estado, May 4, 1896, exp. Ayahualulco.

73 Santiago Trujillo to Alcalde Municipal de Ayahualulco, November 10, 1893, exp. Ayahualulco.

74 Jefe de la Sección de Fomento de Veracruz to Secretario del Gobierno, November 16, 1893, exp. Ayahualulco.

tential ire of villagers, but at the same time they inadvertently ensured that the process would be a lengthy and difficult one. Ultimately, on the ground, the land division process rarely cohered as a state strategy. Rather, it was a contested process, one of tactical give and take by a wide range of actors: municipal, regional, and federal authorities; landowners and local elites; and, not least of all, villagers and surveyors. The various problems arising from their interactions and divergent interests complicated the land division process and impacted the pace at which it proceeded. So too did agrarian custom and practice.

III. PRACTICES: HISTORIES AND CUSTOMS

In his beautifully realized examination of the poetics of spatial consciousness, Gaston Bachelard observed that "space that has been seized upon by the imagination cannot remain indifferent space subject to the measures and estimates of the surveyor. It has been lived in, not in its positivity, but with all the partiality of the imagination."[75] Bachelard's oppositional pairing of the surveyor and the inhabitant, of indifferent space and imagined place, serves as a useful reminder that any analysis of the contentious process of land division cannot be reduced to questions of the quality of the survey or the misdeeds of surveyors. Such technical reductionism, mired solely in an analysis of surveyors' inattentions or regional elites' intentions, misses the profound importance of spatial logic and practices. To villagers, the measure of success was not necessarily how well the surveyor measured. Many villagers perceived the survey through a lens of historical and practical intangibles, not amenable to the measuring chain and the quantifying spirit.

There was, in other words, an epistemological disjuncture between the dictates of transparency and the determinants of daily life. The logic that organized land as a patchwork of permanent parcels—scientifically surveyed and neatly mapped—was illogical in that the legal reality of a plot now bore no necessary or inherent relation to the reality of how the land was used.[76] How, for example, could one survey and map the "properties" of vanilla cultivators who often had rights to vanilla plants grow-

75 Bachelard, *The Poetics of Space*, xxxii.
76 Cronon, *Changes in the Land*, 74.

ing in a multitude of areas and no necessary rights over the land or even to other crops growing upon it? These were dispersed and not amenable to consolidation into a single plot.[77] As Pierre Bourdieu has incisively observed, "[P]ractice has a logic which is not that of logic."[78]

Even in cases where families or individuals had certain rights to bounded plots, permanent parceling proved problematic. For example, villagers in the sierra of Chiconquiaco traditionally sowed what was known as a *rueda*. The rueda, while a bounded unit of cultivatable land, was not, as one contemporary pointed out, "a determined agrarian measure, but an extension of land considered sufficient to provide a regular harvest in order that they do not go without maize during the year."[79] That is, the size of the rueda changed yearly, even seasonally, to account for the vagaries of long-term weather patterns, soil quality, changes in the size of a family, the kind of crop being grown, and so forth. A highly local and contextual measure, it responded to the villagers' need to minimize risks (of hunger and starvation) rather than to maximize profits.[80]

Thus villagers and village authorities often struggled to modify the land division in such a way as to dampen the most egregious effects of the simplifying logic of the survey. The most prominent example is *condueñazgo*, an early modification of the land division process whereby land would be divided into large lots [*grandes lotes*], each one held communally by a determined number of *accionistas* (shareholders).[81] The state government grudgingly legalized condueñazgos in 1874 because of concerns over potential resistance to the land division.[82] Condueñazgo

77 For a detailed study of vanilla cultivation and tenure, see Kourí, "The Business of the Land." For an excellent, sensitive study of how common rights do not map directly onto territory, or even crops, see Peluso, *Rich Land, Poor People*. See also Peluso, "Whose woods are these?"

78 Bourdieu, *Outline of a Theory of Practice*, 109. See also Scott, *Seeing Like a State*, passim.

79 María Rodríguez, *Apuntes sobre el Cantón de Xalapa*, 70.

80 On minimizing risk see Scott, *The Moral Economy of the Peasant*.

81 There is a significant amount of literature on condueñazgo. For good discussions, which do not all necessarily share the perspective given here, see Ducey, "Liberal Theory and Peasant Practice"; Escobar and Gordillo, "Defensa o despojo?"; and Kourí, "The Business of the Land."

82 Ayuntamientos in Veracruz were permitted to divide lands into condueñazgos by legislative decree on December 7, 1874. See Escobar and Gordillo, "Defensa o despojo?" 23–25, and Kourí, "The Business of the Land."

permitted traditional forms of agrarian production to persist by main-
taining a semicommunal form of tenure in which the numerous small
and dispersed plots of cultivation could be maintained under a single
group and the land could lie fallow when necessary.[83] Even forms of condueñazgo could be further contextualized. Municipal officials in Minatitlán received permission to divide their lands into twenty-five large
lots. In drawing up the contract with the surveyor, they required that he
carry out two distinctive surveys of twenty-five large lots each: one of
land along the banks of the river—composed of "savanna, highlands and
lowlands, around the town"—and another of the lands on the island in
the river—composed primarily of "land for cultivation and pasture."[84]

Modifications occurred in other ways. In Acultzingo, Holzinger—
even before he renegotiated his survey with the town merchants—
agreed to divide the communal lands with three distinct surveys: one of
riego land, one of *temporal,* and a survey of land set aside as part of a
ranchería.[85] The ability to contextualize the survey in this way was critical to its successful completion. The *ojos de agua* (natural springs) in
Acultzingo were the lifeblood of agriculture, ensuring the success of
crops such as corn, beans, chickpeas, chile, and barley, and permitting
the continuous cultivation of the lengthy valley between Acultzingo and
Orizaba.[86] Indeed, their importance to the village is imprinted in the
daily imaginary of its inhabitants in its name, which means "spring be-

83 See Ducey, "Indios liberales y tradicionales," 16.

84 See Alcalde Municipal to Jefe Político, September 6, 1883, AGEV, RG, Tierras, Reparto, 1903, exp. Terrenos: Lo relativo a la división y reparto de los comunales del Municipio de Jáltipan, Minatitlan y Copia del contrato celebrado entre
el H. Ayuntamiento de Jáltipan y el Ingeniero Albino G. Bradstreet para la medición y reparto de terrenos comunales, October, 1883, in ibid.

85 "Riego" referred to land that could be irrigated, abutting one of the water
channels. "Temporal" was land where crops could be grown only during the rainy
season.

86 Arroníz, *Ensayo de una historia de Orizaba,* 16–17. Francisco de Ajofrín, a
weary friar who sought refuge from the "insular and aggressive indians" of Acultzingo in 1763, marveled that the hacienda of San Diego had enough water to perpetually irrigate its crops of barley and garbanzo beans. Ajofrín, "Diario del Viaje
que hicimos a Mexico," 2:59–60. Water also cultivated contentious conflicts between the townspeople and hacendados. Just two years before Ajofrín's visit, disputes over water raged between the villagers and don Francisco García Mellado,
who rented lands from the hacienda de San Diego. See AGN, Ramo Tierras, 1761,
vol. 879, exp. 5.

FIGURE 6. The iconographic landscape: Anonymous copy of 1559 map of Acultzingo, 1895. Courtesy of the Mapoteca Manuel Orozco y Berra, Mexico City. Photograph by Carmen H. Piña.

neath a tree," and iconographically accented in the village's 1559 map (figure 6).[87] On Holzinger's land division map (figure 5), the ojos appear as relatively unremarkable rivulets, leaking across the land. In contrast, on this image the channels are veritable rivers of life and they dominate the image. Any land division that would have not ensured each *agraciado* (land recipient) a plot along or within channel access to the banks of one of the five major waterways would have stood little chance of survival.

87 This according to a local historian in the town and based upon a story still told today, that Acultzingo was founded by four caciques who descended from the surrounding sierra to a place where water coursed from beneath a tree. Other villagers in Acultzingo assured me that the name meant "where there is much water." Interviews by the author, May 13, 1999. Both Martin Holzinger and the nineteenth-century *cronista* of Orizaba, Joaquín Arroníz, similarly equated the

As much as they mitigated the effects of the division, neither con-dueñazgo nor other kinds of modifications constituted durable solu-tions. Condueñazgo persisted only until an increased financial and mili-tary capacity permitted the state to break these up into individual lots or until local power holders saw the potential benefits of the permanent titling and codification of parcels, as occurred in Papantla in 1895. More to the point, as much as villagers were able to modify the terms of the division to account for local context, what they could not alter was the process of reification. By fixing property as decontextualized permanent parcels, the land division held the world static, inscribed, and codified in the bureaucratic map. But places changed over time. The example of Acultzingo is instructive: in an era of steadily decreasing rainfall totals, many plots divided as riego land in 1870 were little better than plots of temporal by 1900.[88] Protests in Papantla in the 1890s were partially a re-sult of the fact that children of those granted parcels in the original land division—children who had been too young to receive a plot them-selves—were now finding themselves landless adults.[89]

If the land division could not account for the future, neither would it be beholden to the past. The timeless geometry of the surveyor's grid negated the historical geography of the villager's ground. This became particularly clear once the survey was finished and the process of plot distribution began. At the beginning of the survey, municipal authori-ties posted notices of the impending division, requesting that anyone in the community with rights to land and desiring to participate, sign up in order to be considered an agraciado. The final list of recipients (the padrón) would then be used as the basis for determining how

---

name with water: Arroníz suggested the name meant "where the water turned," while Holzinger thought it derived from one of the ojos of agua named Acoltze. See Arroníz, *Ensayo de una historia de Orizaba*, 490–91, and Holzinger, "Informe." Regardless of the precise meaning of the word, one feature remains constant: water.

88 Not surprisingly, by 1900 plots of riego were worth ten times more than plots of temporal or monte land of similar or equal size. See AMA, Tesorería, leg. 2 (1900–1904), exp. 1–8. On the decreasing rainfall, see Arroníz, *Ensayo de una his-toria de Orizaba*, 44–5. This continues today. Even during years of good rainfall, water from the ojos no longer reaches the riego plots at the easternmost portion of the town.

89 See Ignacio Muñoz to Gobernador Constitucional del Estado de Veracruz, December 17, 1894, exp. Ingenieros, and Kourí, "The Business of the Land."

many lots would be created by the surveyor as well as their relative size.[90] The usual manner in which the actual granting of plots was done was through a *rifa* (or *sorteo*)—literally, a raffle in which lots would be assigned to individuals through the presumably random process of a drawing. Designed to minimize the potential for corruption in the land distribution, the rifa may have seemed the perfect solution, but it had numerous problems. In the first place, regardless of the theoretical presumption that the rifa would resolve questions of corruption in the doling out of land, corruption still persisted. The signatory to a letter directed to the state governor by opponents of the land division in Misantla observed that persons of wealth and influence in the community were overwhelmingly receiving the best quality lots.[91]

Moreover, the problems that inevitably arose from the imposition of an abstracted spatial configuration onto a lived environment plagued the process. While the concept of a rifa in theory may have been attractive, it ultimately faced stiff opposition once villagers became aware of its ramifications. It would appear that on many occasions, villagers were unaware of how the rifa worked and did not understand that the lot they received would not necessarily encompass the land they worked or the area in which they grew their crops. That is, the rifa did not confirm their individual rights to the land they worked and crops they cultivated but merely ensured they received a plot of land somewhere within the village bounds. Consequently, villagers often received plots at significant distances from where they lived and worked. In the case of the former, they were put in the unenviable position of having to either move from where they had settled or remain away from their household for extended periods. In 1886 a number of villagers wrote the state governor complaining that they were to "receive our respective lots in places very distant from the *congregación* [small village or hamlet] and where for reasons of distance we cannot cultivate without sacrificing our local interests in the congregación, abandoning our home and our families."[92] If

90 For a good example of how this worked, see Holzinger, "Informe."

91 Santiago Tinoco y otros vecinos de Misantla to Co. Gobernador del Estado, December 29, 1886, AGEV, RG, Tierras, Reparto, 1872, exp. Misantla [hereafter exp. Misantla], ff. 176–78.

92 Felix Mogollón y otros vecinos de la Congregación de Culebras to Co. Gobernador del Estado, exp. Misantla, ff. 167–68. For similar and repeated complaints, see Francisco Mora y otros vecinos de Misantla to Gobernador del Estado,

forced to abide by the division, they claimed, their congregación would surely disappear as they would have to "form more or less insignificant *ranchos* on our respective possessions, and as a consequence lose the comforts, prerogatives and advantages which a nuclear population provides."[93] Tapping into the state's own rhetoric regarding education, they went on to ask how their children "will be educated and enlightened" if their land were given to other community members. Not surprisingly, they suggested they be given new plots next to their congregación.[94]

Others questioned the loss of labor invested in certain portions of the land, indignantly stating that they did "not believe it just that after having opened up virgin lands at the cost of much sweat and labor" they would now lose that land through a rifa.[95] Villagers stood to lose more than the lands they had cleared. Crops such as vanilla or fruit trees required significant inputs of labor before they began to yield with regularity. Moreover, these crops constituted the patrimony that villagers wished to pass down to their children. In other cases, villagers had built small dwellings on portions of land that they now were to lose. In sum, the rifa ignored both the labor and meanings villagers had individually and collectively invested in and ascribed to the land. No wonder many villagers looked out across the rapidly changing landscape "with true sorrow and great bitterness."[96]

In some instances, villagers asserted the right to alter the land distribution in such a way as to maintain the lands they had worked. An 1889 report to the state governor on a land survey in Santiago Tuxtla ob-

May 13, 1886, exp. Misantla, ff. 169–70; and Ramón Salazar y los demás interesados que no saben firmar to Co. Gobernador del Estado, December 26, 1886, exp. Misantla, ff. 179r–v. This does not, it should be emphasized, mean that all were becoming landless. The primary signatory to the above letter, Felix Mogollón, was doing fairly well in terms of holdings by 1908. He owned lotes in three different parts of Misantla valued at some 12,500 pesos. See Relación de las personas que poseen fincas rústicas en el Municipio, AMM, caja 1910, exp. Presupuesto de gastos del Municipio de Misantla . . . 1910.

93 Felix Mogollón y otros vecinos de la Congregación de Culebras to Co. Gobernador del Estado, exp. Misantla, ff. 167–68.

94 Ibid.

95 José María et al. to Gobernador del Estado de Veracruz, n.d. [ca. late December, 1904], exp. Yecuatla, f. 204 r–v.

96 Santiago Tinoco y otros vecinos de Misantla to Co. Gobernador del Estado, December 29, 1886, exp. Misantla, ff. 176–78.

served that "some agraciados have more or less quantity of land" than others. The author continued that, while such a difference might at first appear unjust, it was the result of the wishes of a number of "agraciados who, having possessed certain parcels from a long-ago time, preferred to conserve these, although they would have less land surface than those lots indiscriminately adjudicated."[97] In Minatitlán the prefect determined that villagers could keep the houses they had built even if they had been granted a different lot in the land distribution as long as they did not assert any permanent rights to those lots.[98] But more common may have been the response of the prefect to the protestations of a group of villagers in the canton of Misantla, protests that drew together questions both of agrarian practice and history. He admitted that cultivators of vanilla had been negatively effected by the land division and distribution because of the "material impossibility of subdividing the lands dedicated to sugar cane or vanilla, which are in general very small, almost insignificant, and *irregularly established.*" He then dismissed the concerns of the cultivators, arguing that little could be done considering "the absolute inconvenience of burdening the agraciados and the Ayuntamiento of this villa with the enormous expenses that would have to be incurred in order to *rectify the maps*, [and] alter and change the possessions already granted."[99] Thus did the map now dictate the future.

The rueda, rights to crops rather than territory, the personal and familial histories of use—the entire imaginative complex of daily life—these all defied easy summation. There were, and are, no terms in classical liberal theory to account for them. Much like a rational choice theorist's model, the land division and its epistemology reduced the reality of human existence to "noise." Modern freehold property ideas were, in one author's felicitous image, akin to a "straitjacket" that simply could not "satisfactorily represent . . . the buzzing complexity and plasticity of customary land tenure practices."[100] Nor could they account for the constant changes and modifications of such practices that occurred in a

97 In *Memoria presentada . . . el 18 de Septiembre de 1890.*

98 Comisión de Gobierno to Gobernador del Estado, June 28, 1886, AGEV, RG, Tierras, Reparto, 1903, exp. Terrenos: Lo relativo a la división y reparto de los comunales del Municipio de Jáltipan, Minatitlan.

99 Jefe Político to Secretario del Gobierno del Estado, January 24, 1887, exp. Misantla, ff. 173–75. My emphases.

100 Scott, *Seeing Like a State*, 300.

changing, lived environment. The issue here was not necessarily (or solely) that of the imposition of private property on to a different tenure structure. As already noted, the "communal system" often easily accommodated forms of personal tenure.[101] Rather, the point is one of the qualitative manner, rather than the absolute fact, of land division: of land as contiguous and *fixed* parcels removed from their temporal, spatial, and social context. If fixity promised a kind of stability for buyers and sellers of land and for state bureaucrats, it could often make life very unstable and insecure for cultivators accustomed to other forms of spatial organization directly related to agrarian practice, production, and custom. Indeed, the very idea that land could be fixed in such a way developed out of a conceptualization of space as a thing distinct from— rather than produced by—the practices of the people that inhabited "it."[102] Property thus became an object itself rather than a relationship.[103]

As a consequence, the state's desire for cartographic transparency meant in reality a complete reconfiguration of the ground. In this sense, the map did not reflect the ground: it reordered it, forcing agrarian practices to conform to its own normative vision. The only reflection the map offered was that of a liberal state's market and bureaucratic logic.

IV. RETHINKING "RESISTANCE"

With this in mind, it would be worth revisiting Governor Teodoro Dehesa's blunt declaration in 1897 that "the clase indígena has always tenaciously opposed the division of communal lands."[104] Attempting

101 It bears reiteration here that the communal system, relations in villages, use rights, and so forth should not be romanticized. Duties, rights, and responsibilities in villages were rarely fairly or evenly distributed and they were often exclusionary and parochial. For studies of customary use-rights and the commons that avoid sentimentalization, see Thompson, "Custom, Law, and Common Right"; Williams, *The Country and the City*; and Jacoby *Crimes against Nature*.

102 Abstract space and lived space thus have their analogues in Marx's conceptualization of, respectively, exchange value and use value. See the classic analysis of Lefebvre, *The Production of Space*. Also useful are Smith, *Uneven Development*; Gregory, *Geographical Imaginations*, esp. 274–77; and Cosgrove, "Prospect, Perspective and the Evolution of the Landscape Idea."

103 Or, as Thompson puts it, a right *of* rather than *to* things. Thompson, "Custom, Law, and Common Right," 135.

104 Quoted from Escobar and Gordillo, "Defensa o Despojo?" 26.

to explain why lands still remained undivided, some seventy years after the first promulgation of a land division law in the state and nearly half a century after the passage of the federal laws of the Reforma, Dehesa's claim reflected two prevalent assumptions among the Liberal and Porfirian establishment: that surveys did not proceed apace because of a unified resistance on the part of an "Indian class"; and that Indians were by their very nature opposed to the land division because of some innate antiliberal, communitarian ethic. Yet a closer look at the actual process of land division suggests that such accusations, with their conflation of ethnic identity and ideology, could hardly account for the lethargic pace at which land divisions proceeded in nineteenth-century Veracruz.

In the first place, not all Indians opposed the land division. An emphasis upon "Indian resistance" reduced largely indigenous villages to unified and homogeneous entities. They were not. There could be, and was, unity within diversity, but the fact remains that villages were at times riven by different kinds of conflict. Within communities some supported the land division, while others did not. The municipal authorities who contracted surveyors often were, after all, Indians. In Acultzingo, Domingo Guzmán, purportedly the only living descendant of the indigenous founders of the community in 1554, served on the municipal council the very year it hired Holzinger to divide the lands. After the land division, he received a special set of lands not subject to the same prescriptions as the standard lots given to others. And while Holzinger never named the merchants who paid him to do the land division and expand its parameters, we should not automatically presume that they were non-Indians or nonmembers of the community. By 1908 a small number of villagers, not just outsiders, had consolidated significant amounts of land.

Moreover, the constant delays in surveys were often consequences of local contestations over the process of implementation rather than simple opposition. As noted, disputes often erupted over *who* should divide the land or *how* it should be done and not necessarily over the fact of land division itself. Dehesa, of course, did not entirely imagine "resistance" when he penned his speech in 1897, but one should not assume that such resistance had always been there or that it was uniform and unchanging. Such a conception merely serves to strengthen notions of a romantic but static tradition and locates villagers outside of history. Outright opposition to the land divisions was not, as Dehesa seemed to suggest, a visceral response conditioned by some primordial antiliberal chromosome buried deep in the Indian's genetic code. The process of implementing the land division left much to be desired in the eyes of many, and resistance to surveys most likely sprung from more historical (and admittedly mundane) roots: a historical memory of error-ridden or incomplete surveys, dubious surveyors, and unsympathetic authorities. Land surveys were riddled with mistakes and problems or were simply not carried out in the way they were supposed to be; surveyors did not fulfill their obligations or were imposed upon villagers by interested federal, regional, or municipal officials at exorbitant cost; and they were often neither trained nor impartial.

Resistance also increased as villagers learned how little control they had over the final form of land distribution and the ways in which parceling would effect their agrarian practices, livelihoods, and existing properties. Holzinger's journal, for example, reveals that opposition by both villagers and village authorities developed at specific points in time *over the course of his survey*, such as when he plotted out the fundo legal. When Holzinger arrived in Acultzingo, he confronted a town "without a well-demarcated fundo legal and only two streets that scarcely deserve the name."[105] Determined to create one, he mapped into existence seven rectilinear streets and some sixty-four blocks of urban plots, "each with the symmetry which Modern Towns require and conforming to the Topography of the land."[106] His actions generated vociferous opposition among villagers and the authorities, who saw the urban plots and homes they had created being rent asunder to conform to Holzinger's aesthetic sensibilities. Even the town priest voiced his opposition, angered as he was by Holzinger's attempts to put a road through his farm abutting the church. Resistance, in sum, was conditioned by historical experiences with the practical implementation of the land division, not a predetermined reaction dictated by ideology.

CONCLUSION

In calculating a new reality, the land division map forecast the demise of another. Once surveyed into existence, the lots Martin Holzinger had

105 Holzinger, "Informe."
106 Ibid.

plotted out on his map quickly assumed a reality on the ground. The plots circulated, or more precisely, they flowed unidirectionally. Common were the kinds of transactions to which villagers Julián Espinoza and Andrés Acevedo were party in 1887. His father having recently died, Espinoza sold his fundo legal plot to Acevedo for the sum of fifteen pesos to cover funeral expenses.[107] The transaction required nothing more than a signature on a small piece of paper, denoting the lot number and its boundaries and the transfer of that piece of paper from one hand to the other. Such transactions were increasingly common, and by 1900 the geometric lots had become standardized commodities, concentrated in the hands of an increasingly small number of villagers and regional elites. From 1902 to 1908, there was a virtual frenzy of land transactions, and by 1918 a full quarter of monte land and nearly a third of all temporal land was held by some twelve individuals.[108] There is a bitter irony in this: Holzinger had lamented in 1872 that the "unfortunate indígenas" of Acultzingo were "owners of everything . . . but possessed nothing." By 1911 most villagers in Acultzingo owned nothing and possessed little. Land had become concentrated in fewer and fewer hands, and, in time, Acultzingo became a hotbed of agrarian radicalism.

Holzinger's map still hangs on the wall of the municipal town hall. As I carried out my research I watched as it was repeatedly pulled down and laid flat on a lone table, to be consulted by buyers and sellers of plots of land. After the pointing and tracing was done, it was returned to its place on the wall: a permanent fixture in, and of, Acultzingo.

107 Documento privado de la venta de un solar del Fundo Legal marcado en el número 118 que otorga el C. Julián Espinoza al C. Andrés Acevedo, July 25, 1887, AMA, Independiente, Presidencia (1866–1949), leg. 1, exp. 5.

108 AMA, Actas de los Lotes de Terrenos de Temporal, 1871.

＊

CHAPTER 4

Situated Knowledges:

The Geographic Exploration
Commission (I)

In his first memoria to the Mexican Congress in 1879, Agustín Díaz, head of the fledgling Comisión Geográfico-Exploradora [CGE], did not mince words: the project of obtaining precise statistical and geographical knowledge through the exploration and mapping of the country was "a business upon which surely, more than any other, the future of the Republic depends."[1] Although it was also a business upon which his own professional future depended, Díaz's statement was not mere self-interested hyperbole. Some of the most prominent figures in the new administration, including President Porfirio Díaz himself and his minister of development, Vicente Riva Palacio, shared his opinions regarding the need for reliable, comprehensive maps of the country and they supported the creation of the CGE in 1877.[2] Both understood the importance of systematic exploring, surveying, and mapping of the country to the consolidation of rule, defense from foreign invasion, and promotion of economic development.

Díaz's statement suggests that states, like empires, are united as much by information as by force.[3] Indeed, the increasing cumulative weight of the Porfirian state can be discerned as much in the hefty tomes of statistical and geographical knowledge produced per annum as it can through reference to the size of its military. In the case of the CGE, knowledge and force intersected in a very immediate and powerful way. Charged with

1 Díaz, *Memoria de la Comisión Geográfico-Exploradora . . . 1878 a 1879*, 3.
2 See Porfirio Díaz, *Informe*. Agustín and Porfirio were not related. For purposes of clarity in this chapter, when referring to Porfirio Díaz I will use "don Porfirio." All references to "Díaz" are to Agustín Díaz.
3 See Richards, introduction to *The Imperial Archive*.

producing and structuring geographic knowledge, military personnel dominated the agency and the needs of the Ministry of War determined the content and forms of their surveys. Cartography and the military, knowledge and force, were mutually reinforcing.[4]

This chapter examines the CGE's first two decades of existence to show how mapping and the military functioned together in the process of Porfirian centralization. In part 1 I examine the rationale behind the formation of the agency, putting particular stress on how the CGE represented a new federal initiative to create a uniform archive of structured knowledge through a coordinated mapping and surveying effort. In part 2 I look at the dominant role played by the military in the agency. In parts 3 and 4 I examine how military surveyors surveyed in order to suggest two things: first, that while the traverse (or route survey) could indeed help create an archive of structured geographical knowledge in the form of scientific maps, it was also technically problematic; and second, its usefulness thus derived from the fact that it generated an extensive staff of military officials with detailed, situated knowledge of the locales and regions they were charged with mapping. This combination of "structured" and "situated" knowledge became a critical cornerstone of Porfirian statecraft.

I. CARTOGRAPHIC ANARCHY

When Porfirio Díaz assumed power in 1876, he appeared determined to create more accurate maps of the country. His determination was in part dictated by personal experience: he had learned first hand, during the war against the French Intervention and his own rebellion of Tuxtepec, of the need for comprehensive and reliable maps of the country.[5] In Jan-

4 On mapping and the military as reinforcing technologies, see Thongchai, *Siam Mapped.*

5 Díaz, *Exposición internacional colombina de Chicago*, 4; Díaz, "Catálogo de los objetos que forman el contingente de la citada Comisión. Exposición internacional de Paris," in ACGE, paq. 162 [hereafter: Díaz, "Catálogo de los objetos"]. [Note: there is no pagination or organization to most of the materials in the ACGE. When available I note the folio or page number.] War invariably seemed to spur geographic and cartographic initiatives in the nineteenth century. For example, geography became a prominent discipline in France following the loss of the Franco-Prussian War in 1870. France's defeat was attributed to a lack of geographical knowledge and flawed maps. See Ross, *The Emergence of Social Space*, 93–94.

uary 1877 he thus ordered his minister of development to appoint a commission to review the state of Mexican cartography.[6] The Cartographic Commission came under the charge of Agustín Díaz, who, after reviewing some five hundred maps, reached a dismal conclusion:

> The anarchy that one sees in this collection with regards to the forms of projectión used, scale, signs, typefaces, characters, etc., make any comparison between them extremely difficult; and the differences in the names of places, in their categorization, and in their absolute as well as relative locations, makes any concordance or coordination between them almost impossible. In the numerical data that connects them to the map one also sees significant divergence and encounters such contradictions as to cause doubt even in those [geographic] positions derived from authorities in whom one had great trust.[7]

Given the premises of the Commission, this was as much a commentary on the state of state formation in 1877 as it was on Mexico's cartographic condition. By way of conclusion, Díaz suggested starting all over by forming a new cartographic agency to create a master carta general, of uniform construction and appearance, that would render a true and accurate statement of geographic reality.

Riva Palacio took Díaz's recommendation to heart and to the president. By the year's end the CGE had been officially created.[8] Riva Palacio appointed Díaz to head the agency. He was an obvious choice. His credentials included a distinguished military career, substantial surveying and map-making experience on the northern border with the Comisión Mexicana de Límites con los Estados Unidos, and a lengthy stint as a professor of topography at his alma mater, the Colegio Militar.[9] His three years of service in the Yucatan as part of a scientific expedition organized under the auspices of Maximilian's imperial regime constituted the one stain on his otherwise venerable career.[10] Ironically, it was a

6 Manero, "Informe de la Sección Primera," 455.

7 Díaz, "Informe sobre el estado actual de la cartografía," 475–78.

8 García Martínez, "La Comisión Geográfico-Exploradora," 489.

9 Sánchez Lamego, "Agustín Díaz, ilustre cartógrafo mexicano."

10 A number of Díaz's reports can be found in documents related to the Comisión Científica de Yucatán in the Centro de Apoyo de Investigaciones Históricas del Yucatán, Mérida. Later hagiographies of Díaz would fail to make mention of

project Díaz undertook while in the Yucatan—a project designed to spatially rationalize Mérida in order to facilitate property taxation and create a comprehensive property cadaster for the offices of the treasury— that assured Riva Palacio that here was a man with the sensibility and experience to put science to work in the service of the state.[11]

The aim of the CGE—to systematically construct a master map of the Republic, at a scale of 1:100,000—made it the most ambitious cartographic project in Mexico to date. In contrast to the pictorial-descriptive maps of García Cubas, respected as they were, the CGE would create maps based not solely on the deductive coordination of existing maps and data but also on actual fieldwork: measuring distances, fixing control points, taking astronomical readings with sophisticated instruments, and traversing the land. The maps would be constructed using uniform methods and the latest scientific techniques, with a coherent cartographic code of symbols and coordinates. The final products would render obsolete existing vernacular, unreliable, and presumably inferior maps of the territory, standardizing their information into a new, unifying structure. The work of the CGE would thus provide a uniform frame-

---

his connections to Maximilian's regime. See for example de P. Piña, "La Comisión Geográfico-Exploradora y la influencia de sus trabajos."

11 Díaz to Vicente Riva Palacio, June 30, 1877, BLAC, Vicente Riva Palacio Papers, 1790–1896 [hereafter VRP], W-185, folder 2, letter 249. See also the collected materials in the folder "Agustín Díaz, Proyecto de nomenclatura de calles, presentado al Sr. Secretario de Fomento, Vicente Riva Palacio," ACGE, carpeta 106. The crux of the project was twofold: first, to construct a series of detailed cadastral maps that would permit treasury officials to "judge the relative values of each property with a single glance" and to "discover" and duly tax property hidden from bureaucratic view; second, to rationalize the numbering system of the city in such a way as to make it connotative, such that "on a dark night it would be enough to read two numbers (consecutive or not) in order to know the cardinal points, the precise place where one was, and even the distance from one's destination." Díaz in fact envisioned Mérida as a testing ground for a nationwide project with great bureaucratic potential: a bureaucrat from the Ministry of Finance would be able to move "from one city to the next without even having to ask his predecessor where each finca is or where to find it in the property registration books." If such a project were carried out in Mexico City, Díaz suggested, the increase in tax revenues stemming from the discovery of hidden or entailed property would be more than enough to cover the costs of the project and state governments would no doubt soon follow suit, extending the presumed benefits of the system nationwide. See "Proyecto de nomenclatura."

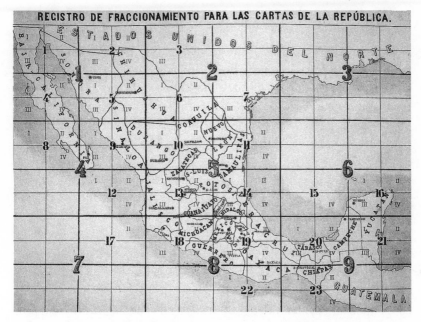

FIGURE 7. Fractions: Comisión Geográfico-Exploradora, *Registro de fraccionamiento para las cartas de la República*, 1877. Courtesy of the Mapoteca Manuel Orozco y Berra, Mexico City. Photograph by Carmen H. Piña.

work around which to begin to envision and codify an array of territorial information; and it would establish a cartographic template for reconciling land disputes, multiple place-names, topographic uncertainties, and an entire, generalized state of cartographic anarchy.

The project hung upon a geometrical, rather than political, framework whereby Mexico's tapering geography was overlaid with a rectangular grid of nine parcels. Each of these grids was then divided into four equal parts and then split into four quadrants again (figure 7). The approach, Díaz argued, would permit the agency to easily correct and supplement the master map of Mexico, building toward an ever more perfect and complete rendering of the territory.[12] The CGE's productions would set the standard and frame within which all subsequent maps would be contained. They would constitute the structural scaffolding for continual incorporation and organization of spatial information and the production and circulation of geographic knowledge.

12 Díaz, *Atlas General de la República Mexicana*, July 1877, MOB, CG, CGE.

This "cartographic ideal"—of a comprehensive and definitive archive of geographical knowledge—was not solely an end in itself.[13] It held great cultural, political, and economic promise for the developing state. The maps, for example, would "popularize the geography of the country."[14] Díaz wanted to create affordable versions of the CGE maps for use in municipal, state, and federal offices; in schools; and by the general public. Such maps would be specifically tailored for the audiences in question. For example, rather than using the relatively new cartographic language of contour lines to show topography and relative elevation, these maps would use the older system of hatch marks, which, while less precise, had the effect of producing a perceived shadow, giving the flat map a sense of tridimensionality.

The maps would also help create a more reliable base for the practicalities of administration. Díaz had observed that "[e]veryone recognizes that the principal cause of administrative errors [is] a lack of knowledge of the areas [of the country] . . . with its policies not being based upon an exact knowledge of the facts, no law, no government provision, not even a simple juridical decision, will be enforceable."[15] No longer would the government be dependent upon a sparse selection of conflicting and error-ridden maps, populated with incommensurable geographic coordinates and uncertain place-names. Nor would it be forced to rely solely upon the irregular submission of dubious information from local authorities, which foiled any attempts to rationally reconfigure the administrative landscape. Finally, the maps also had great military potential. Indeed, the CGE became the primary agency creating general maps for military operations such that "with a simple look the minister [of war] could alter the position [of the troops] or dictate orders for military operations."[16]

II. "A PRACTICAL SCHOOL"

On May 5, 1878, the entire staff of the CGE left Mexico City for Puebla to begin their work. Despite the heavy symbolism surrounding the date

13 "Cartographic ideal" comes from Edney, *Mapping an Empire*.

14 Díaz, "Informe sobre el estado actual de la cartografía." See also Manero, *Documentos interesantes*, 114.

15 Díaz, *Memoria de la Comisión Geográfico-Exploradora . . . 1878 a 1879*, 8.

16 Justo Alvarez, "Departamento de Ingenieros," 165.

and place, it was not an auspicious beginning. The crew totaled a mere eight: Agustín Díaz; his close friend from the Colegio Militar, Julio Alvarado; an assistant; and an escort of five *rurales*.[17] That the rurales outnumbered the engineers must not have inspired much confidence in any onlookers. Nor would the five old mules, cast-offs from the First Artillery Brigade, which carried their meager supplies, minimal gear, and a few old instruments: a barely functioning theodolite (to measure angles for triangulation), one sextant (to determine astronomical locations), one chronometer (for time and thus longitude), two field compasses (for direction), and three *taquiámetros* (for distance).[18] Díaz looked back upon that first month and found little more profound to say than simply: "We began by beginning."[19]

What went wrong? After all, only five months earlier Congress had unanimously supported the creation of the agency and had granted it a staff of nine (excluding the rurales and the mules), which, although still quite small, would have been enough to permit simultaneous surveys.[20] It had the full support of both Riva Palacio and Manuel González—ministers of development and war respectively—who shared jurisdiction over the institution, as well as of the president of the Republic himself.[21] With such an array of supporters, why did the CGE, when it left town that day, appear as little more than a small, ragged detail?

The primary problem the CGE confronted was its relative importance during a time of fiscal uncertainty.[22] It was easy enough to raise one's hand in approval of a new commission important to a new president but quite another to grant it a substantial portion of meager funds. Whatever the merits of the project on paper (and they were numerous), legis-

17 See Díaz, *Exposición internacional colombina de Chicago,* 5; Díaz, *Memoria de la Comisión Geográfico-Exploradora . . . 1878 a 1879,* 3–4; Alvarado, "Informe de la Comisión Geográfico-Exploradora," 299–301; and Duclos Salinas, *The Riches of Mexico,* 98. Rurales refers to members of the Cuerpo Rural, Porfirio Díaz's notorious rural constabulary.

18 Díaz, *Exposición internacional colombina de Chicago,* 5; Alvarado, "Informe de la Comisión Geográfico-Exploradora," 299.

19 Quoted from García Martínez, "La Comisión Geográfico-Exploradora," 494.

20 Díaz, *Memoria de la Comisión Geográfico-Exploradora . . . 1878 a 1879,* 3–4.

21 On Porfirio Díaz's support, see Díaz, *Informe.*

22 On financial shortfalls during the period see *Memoria de Hacienda y Crédito Público . . . 1 de julio de 1878 al 30 de junio de 1879.*

lators were not predisposed to finance a type of work that would not yield visible results for a number of years, particularly with a nearly empty treasury. They vacillated between seeing the project as luxury or necessity.[23] In some respects, any mapping project of such magnitude would appear underfunded, but certainly the CGE had hoped to begin with something more than a dismal 8,500 pesos for fiscal year 1877–78.[24]

The disparity between fantasy and capacity severely circumscribed Díaz's options on how to proceed. The financial constraints limited his technical options and required he achieve immediate results in order to prove the efficacy and value of the agency. In other words, Díaz's initial decisions with respect to the agency revolved not necessarily around how best to fulfill its objectives but around how to ensure the survival of the fledgling institution he now directed.[25] His decision regarding where to begin the fieldwork is demonstrative of these concerns. With such a minimal staff it was impossible for the agency to carry out more than one survey at a time, and Díaz had to decide where best to focus his efforts. He did not deliberate long before opting for the eastern portion of the country—specifically Puebla, Tlaxcala, and Veracruz—and the small staff of the CGE spent their first three years mapping the city of Puebla and its surroundings. Díaz's choice was the obvious one. In the first place, the entire region of central Puebla and Veracruz constituted the foundational geography for a developing nationalist historiography and as such promised potential popular diffusion of the CGE's work.[26] Indeed, in 1881, when Riva Palacio agreed to write a history of the war against the French Intervention and empire, Díaz was able to provide him a variety of geographic and topographic data related to fields of battle and the war, which further underscored the utility of the CGE's work to the Congress.[27] Pragmatic considerations were just as important: the railroad connecting Puebla to Mexico City allowed Díaz to shuttle be-

23 Díaz, *Memoria de la Comisión Geográfico-Exploradora . . . 1878 a 1879*, 15; and Díaz, "Catálogo de los objetos."

24 García Martínez, "La Comisión Geográfico-Exploradora," 491n15.

25 See Díaz, "Informe de la Comisión Geográfico-Exploradora," 71–72; Alvarado, *Comisión Geográfico-Exploradora . . . París en 1900*, 9–14.

26 Alvarado, "Informe de la Comisión Geográfico-Exploradora," 299.

27 Díaz to Vicente Riva Palacio, February 18, 1881, BLAC, VRP, W-187, folder 8, letter 30.

tween the survey sites and the capital with relative ease. Such quick access was key: the agency's precarious financial situation demanded that Díaz be on hand in the city to deal with any potential issues regarding funding and support.[28] Moreover, the area was the region of most administrative, political, military, and industrial importance to the new regime.[29] The first railroad connecting the busy port of Veracruz to Mexico City had been completed only four years earlier; the important cities of Puebla, Córdoba, and Orizaba all lay on its route; and the port of Veracruz still constituted the primary place of entry for travelers and goods coming to Mexico.[30] Invading armies also entered through Veracruz, as Díaz could personally attest, and military concerns likewise influenced his decision. The comprehensive mapping of the access routes to the mesa central, for example, was a project generally considered indispensable to the defense of the nation from foreign invasions.[31] In sum, the seemingly symbolic gesture of departing for Puebla on the anniversary of the battle of Cinco de Mayo obscured the fact that it was in large part a strategic choice based upon pragmatic considerations about how best to proceed under quite dismal conditions.

By the end of 1881, when the CGE moved its base of operations to Xalapa, Veracruz, things had improved. The Commission's financial situation, although still poor relative to the magnitude of the project, was certainly more secure than it had been the previous three years. Between fiscal years 1880–1881 and 1884–1885, the CGE's funding from the Ministry of Development more than doubled.[32] During the same period, the CGE established its own map production offices as well as a small lithography and photography studio; created a department of natural history that collected plants, animals, and artifacts for display in the CGE's newly built Natural History Museum in Tacubaya; and published the first se-

28 Díaz, *Memoria de la Comisión Geográfico-Exploradora . . . 1878 a 1879*, 14. His concerns were not unfounded: only two months after Díaz and his staff left Mexico City, the new Congress attempted to reduce the Commission's funding. They failed. See Manero, "Memoria de la Comisión Geográfico-Exploradora," 2.

29 Díaz, *Memoria de la Comisión Geográfico-Exploradora . . . 1878 a 1879*, 14.

30 On the importance of Puebla, see Manero, "Memoria de la Comisión Geográfico-Exploradora," 2.

31 Díaz, "Informe de la Comisión Geográfico-Exploradora," 1:75–76.

32 *Memoria de la Secretaría de Hacienda . . . 1880 a 1881*, 110–12; and *Memoria de la Secretaría de Hacienda . . . 1884 a 1885*, 132–34.

ries of completed maps for public consumption and bureaucratic distribution.[33]

While an improvement in the state of the treasury made much of this possible, so too did the Ministry of War, which gave the CGE what it needed most: personnel. Díaz had lobbied from the beginning for close military involvement in the agency, arguing that military participation would limit the expenses of the project, combine military analysis into the surveys, and create an entire corps of military engineers with practical exploring and surveying experience who could serve as guides and explorers for future operations.[34] His arguments, as well as the role of military engineers in Manuel González's campaign against a rebellion in Tepic, convinced the War Ministry to expand the CGE's staff by creating a Cuerpo Especial del Estado Mayor (Special Corps of the General Staff).[35] Composed of recent graduates of the Colegio Militar as well as a number of more seasoned officers, the Cuerpo Especial was intended to be a military-scientific commission, assigned with creating "in times of peace the elements that ensure success in times of war": namely, the topographic and geographic maps necessary to any military operation, as well as a carta general.[36]

Paid from the budget of the War Ministry, the new personnel substantially increased the CGE's field staff without burdening it with additional expenses. Simultaneous surveys were now possible, and Díaz could even order a detachment to begin working in Tamaulipas. The impact of the increased personnel can be emphasized with one simple statistic: the total number of kilometers surveyed in fiscal year 1881 surpassed the previous three years combined.[37] Just as interesting, however, is that, by 1882, the entire staff of field personnel in the CGE were drawn

33 Agustín Díaz to Porfirio Díaz, May 5, 1886, CPD, leg. 11, caja 11, doc. 5036; García Martínez, "La Comisión Geográfico-Exploradora," 501 and appendix; and Rosendo Sandoval, "Memorandum relativo al taller de fotografía de la Comisión Geográfico-Exploradora," MOB, CG, CGE, no. 6282.

34 Díaz, "Informe sobre el estado actual de la cartografía," 480.

35 See Díaz, "Informe de la Comisión Geográfico-Exploradora," 71–72. For a list of its staff, see Montesinos, "Relación del personal," 639.

36 *Memoria que el Secretario de Estado y del despacho de Guerra y Marina presenta . . . 1 de julio de 1886 a 30 de junio de 1890*, 9–10. See also *Memoria de Guerra y Marina, 1877–1881*, 10; and Díaz Rivero, *Estudio preliminar*, 49–50.

37 Díaz, "Informe de la Comisión Geográfico-Exploradora," 77.

from the ranks of the military.[38] The dominance of the military would persist in the coming years; even with some civilian increases, in early 1891 minister of war Pedro Hinojosa wrote that the CGE was "composed for the most part of chiefs and officials of the General Staff."[39] Those not a part of the General Staff were ex-military men like Díaz and his assistant, Alvarado: men who had seen extensive military service and worked at the Colegio Militar as professors at the time of their appointments to the CGE. Moreover, Díaz structured the agency as a military outfit, giving the few civilian engineers in the office of map preparation a military rank. The nature of the work, he argued, demanded "the rigorous discipline which can only be achieved through military organization."[40] By the time the CGE arrived in Xalapa, there was no clear differentiation between the military mapping operations of the Ministry of War and those of the CGE.[41]

This is not surprising. Don Porfirio's regime during the first two decades of existence was predominantly a military one. In 1885 three-quarters of the state governors were military generals, and the military constituted the formal institution for the immediate assertion of control.[42] Over the next decade the CGE functioned as "a practical school in which officers complete and perfect the theoretical knowledge they have acquired in the Colegio."[43] Young officers about to graduate from the Colegio Militar as well as established junior officers in the engineering and reconnaissance corps considered an appointment to the CGE a choice position. Their interest was understandable: while certainly not easy

38 Montesinos, "Relación del personal"; *Memoria que el Secretario de Estado y del despacho de Guerra y Marina . . . 1 de julio de 1886 a 30 de junio de 1890*, 82; Porfirio Díaz to Alejandro Prieto, June 8, 1888, CPD, leg. 13, caja 13, doc. 6014; and Antonio Pérez Marín to Porfirio Díaz, May 15, 1887, CPD, leg. 12, caja 9, doc. 4251. See also Craib, "State Fixations, Fugitive Landscapes," 184, table 4.1.

39 *Memoria que el Secretario de Estado y del despacho de Guerra y . . . 1 de julio de 1886 a 30 de junio de 1890*, 9–10. See also Porfirio Díaz, *Informe*.

40 Díaz, "Catálogo de los objetos." On giving civilians military ranks, see García Martínez, "La Comisión Geográfico-Exploradora," 488.

41 In public discourse the military foundations of the CGE tended to be downplayed. Adolfo Duclos Salinas, for example, suggested that civil engineers, "aided by a military staff," comprised the majority of the CGE. This was a substantial reversal of what, in fact, the CGE had become. Duclos Salinas, *The Riches of Mexico*, 97.

42 Knight, *The Mexican Revolution*, 1:17.

43 Agustín Díaz to Porfirio Díaz, January 24, 1889, CPD, leg. 14, caja 1, doc. 230.

work and replete with their own dangers and rigors, exploration and surveying were preferable to the grind of regular military service. One avoided, for the most part, the dangers of battle or of being sent to distant outposts in the arid sierras of the far north or the steamy expanse of the Yucatan peninsula. Conducting the gentlemanly work of scientific observation in nearby Puebla and Veracruz must have been quite attractive by comparison. Exploring and surveying also offered intermittent opportunities to increase one's income. A number of CGE surveyors spent a good portion of their time surveying communal lands in Veracruz, for which they could earn substantially more than the sixty-five pesos per month paid to the lowest ranking officers.[44] Furthermore, who knew what one might discover in the field? In 1888, Díaz and a number of his staff stumbled upon a number of mines in a remote canyon north of Xalapa. They promptly mapped the canyon and then staked a claim to the mines (figure 8).[45] Finally, for military personnel from the areas in which the CGE operated, such as Veracruz, employment promised the luxury of working close to their families.[46]

Not surprisingly, intense competition existed for appointments to the CGE. Potential candidates, as well as their families, flooded don Porfirio with personal requests for work in the agency. They understood what one petitioner stated explicitly in a letter to the president in 1885: "[A] single recommendation from you to the director of the Comisión Geográfica [*sic*], Sr. D. Agustín Díaz, would be enough for me to receive employment."[47] Indeed, most roads to the CGE began with don Porfirio. Entrance to the Colegio Militar, a prerequisite for a commission in the

44 It is unclear whether contracting one's individual services to municipios to divide communal lands was encouraged, prohibited, or merely tolerated by Díaz and/or the minister of war. In certain instances—namely, in the Yaqui and Mayo region of Sonora and in Papantla, Veracruz—land divisions were conducted entirely by CGE personnel at the specific request of don Porfirio. On pay scales, see CFD, fondo 621, carpeta 1, leg. 0005, doc. 1, and Secretaría de Guerra y Marina, Mexico, 18 de Noviembre de 1892, in FCP.

45 Croquis de la situación de Las Minas denunciadas en Agosto de 1888, por el Ing.o A. Díaz en representación propia y en la de sus socios Juan B. Laurencio, Rosendo Sandoval, Francisco Ramírez y Mariano Ramírez, MOB, CGV, Varilla 7, no. 6418.

46 See, for example, Rodrigo Elizalde to Porfirio Díaz, August 5, 1889, CPD, leg. 14, caja 21, doc. 10440.

47 J. B. Súarez to Porfirio Díaz, January 9, 1886, CPD, leg. 11, caja 2, doc. 912.

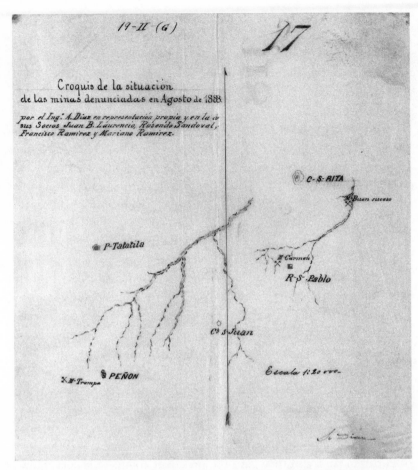

FIGURE 8. A staked claim: Agustín Díaz, *Croquis de la situación de las minas*, 1888. Courtesy of the Mapoteca Manuel Orozco y Berra, Mexico City. Photograph by Carmen H. Piña.

CGE, required don Porfirio's personal approval because of the inordinate number of applicants.[48] Solicitations came from all quarters, from well-placed friends and colleagues in the government (such as Agustín Díaz, who requested that his nephew be admitted to the Colegio) to relatively poor students and their families (such as Luis B. Ulloa).[49] Ulloa contacted the president in 1886 because, despite his perfect scores on his math and French exams, he had not been able to get a scholarship to pursue his studies in engineering. Don Porfirio had Ulloa admitted to the Colegio Militar, from which he subsequently advanced to the CGE. Ulloa would spend the next decades working for the CGE in Veracruz on some of the most sensitive surveying and mapping expeditions. Don Porfirio also ensured appointments for two of his nephews, Félix Díaz and Ignacio Muñoz, when they completed their course work at the Colegio.[50] When they departed for Xalapa they carried a personal letter in hand from their uncle to Agustín Díaz. "The natural interest I have in their fate," don Porfirio wrote, "made me put them at your side because then I am certain they will be successful in their studies."[51] Less than a week after his appointment, Muñoz petitioned his uncle to appoint Rafael Rosas (a close friend from the Colegio), to a position in the CGE. Soon after, the two friends were reunited in Veracruz, where they would play profound roles in the political and social future of the state.[52]

Porfirio's direct interventions served him well, creating a loyal contingent of military surveyors. Over the next two decades, the fortunes of this tight-knit group of CGE surveyors would rise and fall on the tides of Porfirian rule, and they gave the president complete loyalty. However, as the following pages suggest, they provided the administration with something just as valuable as loyalty: an intimate and detailed view of the lands they traversed.

48 Agustín Díaz to Porfirio Díaz, December 29, 1885, CPD, leg. 10, caja 25, doc. 12397.

49 See ibid; Porfirio Díaz to Agustín Díaz, n.d., CPD, leg. 11, caja 1, doc. 387; Luis B. Ulloa to Porfirio Díaz, November 27, 1886, CPD, leg. 11, caja 25, doc. 12111.

50 See General Rivero to Teniente de Ingenieros Félix Díaz, January 4, 1889, CFD, fondo 621, carpeta 1, leg. 0001.

51 Porfirio Díaz to Agustín Díaz, January 17, 1889, CPD, leg. 14, caja 1, docs. 231–32.

52 Ignacio Muñoz to Porfirio Díaz, January 26, 1889, CPD, leg. 14, caja 1, doc. 503.

III. THE CONTINGENCIES OF KNOWLEDGE

By 1885 military surveyors working for the CGE traversed the eastern landscape in increasing numbers. Díaz oversaw a well-coordinated surveying effort crisscrossing the coastal plains of Veracruz and Tamaulipas, the peaks and valleys of the Sierra Madre Oriental, and the plateau of Puebla and Tlaxcala. At his office in central Xalapa he received the numerous route maps, sketches, and reports of men such as Francisco Canóvas, Rafael Rosas, Victoriano Huerta, and Félix Díaz, who, through traverse surveys, brought the terrain into view.

In a traverse survey, the surveyor moved across the terrain, measuring distance and fixing his position and geographic points along his route with a perambulator, a compass, and astronomic observations.[53] Over the course of each day's travels, he recorded direction, distance, and off-route features into a journal or onto a specially designed sheet for topographic sketches. This data would then be forwarded to office personnel who produced an initial *itinerario*, a map plotting the surveyor's route. After comparison with other itinerarios and with existing calculations, a final "cleaned-up" [*itinerario limpio*] sketch was produced (figure 9). These itinerarios, along with the information and geographic positions provided by the surveyor, would then be linked with other traverses; secondary traverse surveys, as well as detailed examination of existing local topographic and property maps, would help fill in the spaces between the route surveys.[54] Surveyors were given explicit instructions with regards to their itinerarios: how to write on them (from left to right); which names (indigenous, ecclesiastical, common, etc.) to give priority; where to begin and end a route (with population centers); how to distinguish roads according to political and military importance

53 A perambulator [*rueda perambulador*] was a long pole with a wheel attached that was rolled over the route traversed and that recorded the revolutions of the wheel with a known circumference. See Francisco Díaz Rivero to Agustín Díaz, December 22, 1881, in "Varios derroteros e itinerarios con alturas barométricas seguidos por Ingenieros de confianza," ACGE, carpeta 106. For an incisive examination of the traverse survey, see Burnett, *Masters of All They Surveyed*, esp. chap. 3; and Edney, *Mapping an Empire*, 91–96.

54 Díaz describes the process in "Informe sobre el estado actual de la cartografía," 478; Díaz, *Exposición internacional colombina de Chicago*, 7–8. See also Gama, "Consideraciones acerca de la cartografía en Mexico," 1:399.

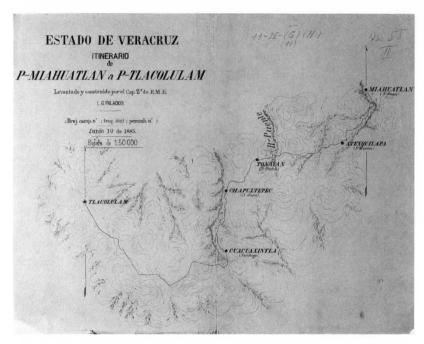

FIGURE 9. The itinerario limpio: Luis G. Palacios, *Itinerario de Miahuatlán a Tlacolulam*, 1885. Courtesy of the Mapoteca Manuel Orozco y Berra, Mexico City. Photograph by Carmen H. Piña.

(with color codes); and how to mark political limits and existing boundary markers.[55] Such codes trained the surveyor in the manner in which he could represent his own experience. While primarily an attempt to organize and standardize collected data prior to submission, these protocols also served a more rhetorical end: they erased the presence of the surveyor, endowing the traverse with an objective quality by eliding the idiosyncratic and subjective nature of the individual's passage. Traverse surveyors did not, as their finished itineraries would suggest, move confidently between two fixed points; as I suggest below, they moved tentatively, occasionally wandering or drifting, as they attempted to get their bearings and negotiate literal and figurative roadblocks. Such wanderings were hardly consistent with the CGE's efforts to build a scien-

55 Cristobal F. Alvarez, "Instrucciones provisionales relativas a la construcción y dibujo de itinerarios," December 31, 1885, ACGE, exp. 19. See also Ignacio Molina, "Informe del Jefe del Departamento de Cartografía," 1:119.

tifically (and rhetorically) structured space of perfected geographical knowledge.

Indeed, neither was the traverse survey itself. The best survey technique for creating an authoritatively structured space, because it offered a "technological fix" for the various errors that permeated the pursuit of the perfect mathematically structured geographic representation, was the trigonometric survey.[56] Trigonometric surveys, however, were extremely expensive and time-consuming, a combination that led Díaz to dismiss them as a viable option.[57] The traverse thus became, by default, the practical means by which to comprehensively map the country. But numerous problems plagued the traverse: its authority relied upon the astronomical observation of the latitude and longitude of a few important places, observations that were not all that certain; each determination of latitude and longitude was independent of the others, such that correction for error was not easy, foiling any attempt at scientific consistency.[58] Compasses and perambulators, moreover, were not particularly precise. Finally, the traverse depended upon the subjective experience of the surveyor himself and the conditions in which he worked. Most errors in surveys arose from errors in measurement—that is, in fieldwork. Traverse surveys were all about fieldwork, and the method of compilation was thus at the mercy of the multitude of points located and distances measured by the surveyors, who operated in difficult conditions.

CGE surveyors confronted serious problems in their attempts to create structured cartographic knowledge. It is an obvious point but one that bears repeating: surveyors did not encounter a stable geography, such as they were charged with providing, but a fugitive, inhabited, and very material world. Their traversals were—in contrast to the maps that resulted from them—fluid and dynamic exercises.[59] In their explorations in coastal and alpine Veracruz, CGE personnel faced an unforgiving landscape and climate, both of which maddeningly slowed the pace of exploration and surveying. On his survey between Huatusco and Cór-

56 I have avoided for the most part any detailed discussion of the technical dimensions of the survey process. For an excellent examination of the workings of the trigonometric survey, see Edney, *Mapping an Empire*.

57 Alvarado, *Comisión Geográfico-Exploradora . . . París en 1900*, 9–14.

58 The following draws from Burnett, *Masters of All They Surveyed*, chapter 3; and Edney, *Mapping an Empire*, especially 91–96.

59 For elaboration on this theme, see Burnett, *Masters of All They Surveyed*.

doba, Alvarado griped that the ruggedness of the "road" on which they traveled limited the number of itineraries they could run and the use of the perambulator, while personnel spent the majority of their time handling the mules to ensure that the instruments carefully packed in their saddle-bags were not destroyed.[60] Regardless, both of Alvarado's chronometers sustained significant damage. Alvarado's traverse went from bad to worse when, after having discovered his instruments damaged, he then had to hasten to Tehuacán to seek medical help for his senior assistant, who had sustained a severe head injury after being thrown from his horse.[61]

The climate in Veracruz affected both the time it took to carry out surveys and the quality of the surveys themselves. Illness plagued the Commission during its stint in coastal Veracruz, a region considered to have a climate so unhealthy that CGE surveyors could only work there two months out of the year.[62] In 1888, while surveying along the coast, two officers in Alvarado's crew died, and he reported that at any given moment some 60 percent of the field staff were intermittently ill from fevers, dysentery, or malaria.[63] While less intense, illness also plagued inland and alpine surveys.[64] Meanwhile, the rains and winds caused by the nortes that saturated central and northern Veracruz from November to March, as well as the torrents of the monsoon season from July to September, left only a small window of opportunity to carry out exploring and survey work with cooperative weather. In the early spring of 1883 the surveys of Agustín Díaz's nephew, Francisco Díaz Rivero, in northern Veracruz were repeatedly delayed as he and his crew sought refuge in ranchos and pueblos from the steady heavy rains.[65]

Weather also affected the results of surveys. The rapid changes in

60 Alvarado to Agustín Díaz, August 15, 1883, ACGE, carpeta 112.

61 Ibid.

62 "Memorandum: Sobre la Carta particular del Estado de Veracruz levantada por la Comisión Geográfico-Exploradora," n.d. [ca. 1906], ACGE, exp. 4, Folleto de Veracruz.

63 Alvarado to Manuel Fernández Leal, February 21, 1888, ACGE, paq. 162.

64 Díaz Rivero, "Memoria de los trabajos hechos por el Capitán 1o. de EME, Francisco Díaz Rivero, sobre levantamiento topográfico del Dto. de Huachinango, Edo. de Puebla, Mayo, 1883," ACGE, carpeta 112 [hereafter: Díaz Rivero, "Memoria de los trabajos"].

65 Díaz, "Informe de la Comisión Geográfico-Exploradora," 74; Díaz Rivero, "Memoria de los trabajos."

pressure that accompanied a weather front impacted the surveyor's ability to take accurate and consistent measurements of altitude.[66] Alvarado's repeated attempts, on his survey between Xalapa and Tehuacán,
to take astronomical readings outside of Perote were frustrated by the
dense cloud cover. When he tried a few days later, a heavy wind repeatedly blew out his lamp and stirred up a dust so thick he could barely
make out any stars at all.[67] Consequently, he based much of his resulting
itinerario on assumptions and relatively few (and poor) observations.
The results could be profound: if "the value of traverse surveys hung
on the precision, accuracy and number of their fixed points," then the
entire enterprise was threatened by the inability to reliably fix those
points.[68]

Surveyors traversed a human, as well as physical, landscape, one no
easier to predict or control. Both landowners and campesinos were particularly wary of the sight of military engineers with land-measuring
instruments, accompanied by a military escort, coming into view on the
horizon. Landowners thought surveyors to be fiscal agents, an understandable suspicion.[69] Pronouncements regarding the need for cadastral
surveys and the creation of a property register rolled off the tongues of
legislators with threatening regularity in the last three decades of the
century. In 1871, after a speech by governor Francisco Hernández y
Hernández lamenting the lack of a property cadaster and the existence
of "hidden" property, the government of Veracruz created a *catastro* section.[70] Over the next three decades, the veracruzano legislature issued
numerous decrees regarding the creation of a statewide property cadastre, one of which compelled landowners to witness the survey process
and dictated severe penalties for any form of resistance.[71] Landowners'

66 Díaz Rivero, "Memoria de los trabajos."

67 Julio Alvarado to Agustín Díaz, August 15, 1883, ACGE, carpeta 112.

68 Burnett, *Masters of All They Surveyed*, 91.

69 See Díaz, *Exposición internacional colombina de Chicago*, 5; and Díaz, "Catálogo de los objetos."

70 Decreto 14, CLEV—*1871*, 439; "Memoria presentada al H. Congreso del Estado de Veracruz Llave, por su Gobernador Constitucional el C. Francisco Hernández y Hernández, el día 13 de marzo de 1869," in Blázquez Domínguez, ed., *Estado de Veracruz*, 2:655.

71 See Decreto 3, January 10, 1887, CLEV—*1887*, 103; and Decreto 31: Ley
para la formación del Catastro parcelario de la propiedad rústica del Estado,
CLEV—*1891*, 140–49.

concerns also arose from their mistrust of surveyors. As noted previously, relatively few professional land surveyors operated in Mexico until the turn of the century. As late as 1889 the governor of Tamaulipas introduced a law to protect hacendados and rancheros "from the ineptitude or bad faith of *Agrimensores* or topographic engineers charged with measuring lands, who always proceed in their operations with most irregularity, without taking the slightest care not to commit errors nor much less to correct them once committed."[72]

As a result of such suspicion, landowners frequently rebuffed attempts by CGE surveyors to consult their property maps. Díaz deemed these maps to be superior in terms of topographic detail to those that could be made by viewing the topography from a route survey.[73] Landowners, however, remained unconvinced that the surveyors wished only to gather topographic detail to fill in the blank spaces on their own maps. In 1884 Díaz pleaded with his good friend Vicente Riva Palacio to convince his propertied colleagues and friends to provide the CGE with maps of their respective holdings. In order to ensure that they would even consider the proposition, Díaz promised Riva Palacio that the CGE would not use the maps for cadastral purposes nor to delineate the boundaries of the properties.[74]

Rural villagers were no less concerned than their propertied counterparts. Villagers interpreted military men with instruments, marking and observing the land, as obvious precursors to either "dispossession or the imposition of new taxes."[75] Whether or not they could distinguish between a property survey (designed to bound a given property) and a military traverse survey (designed to explore and map the land) was irrelevant. Military surveyors in the employ of the CGE carried out both traverse surveys and communal land divisions, the latter at considerable expense to the pueblos in question. That they often neglected to complete the survey as contracted increasingly alienated villagers from the property survey process.

72 Alejandro Prieto to Porfirio Díaz, March 7, 1889, CPD, leg. 14, caja 6, doc. 2537.

73 Díaz, "Informe de la Comisión Geográfico-Exploradora," 75.

74 Agustín Díaz to Vicente Riva Palacio, November 20, 1884, BLAC, VRP, W-188, folder 3, letter 449.

75 See Díaz, *Exposición internacional colombina de Chicago*, 5; and Díaz, "Catálogo de los objetos."

This in part helps explain the varied acts of resistance surveyors encountered in the field. Yet violent encounters were not necessarily the norm. Local resistance tended to be furtive and aimed at the tools of information gathering and knowledge production rather than at individuals. Villagers had a clear enough idea of how the survey worked to know that its potential effects could be undermined without resorting to outright confrontation, which would expose them to retaliation. Thus, in lieu of running the surveyor out of town, they often destroyed station points, pulled up posts, and altered the locations of landmarks.[76]

Villagers also took advantage of the state's relative blindness. For example, they were often hired as guides to help surveyors explore and survey better: they knew the quickest ways to move through the land, the best places for good overviews, the safest roads to travel, the best times of day for clear weather; and—of particular interest—they knew the local names of places and land features. Place-names had repeatedly complicated the state's vision. Little progress had been made in the decades since Romero and García Cubas had penned their respective articles arguing that the federal government assume tight control over the arbitrary changing of place-names.[77] Díaz lamented that the federal government had thus far chosen to ignore the suggestions proffered by these "respectable persons of the Sociedad de Geografía y Estadística."[78] As a result, he continued, not only was the reconstruction of Cortés's route still at issue; so too was the very cartographic enterprise of the CGE. This "ancient and abominable custom," he swore, made an already Herculean task that much more difficult as repeated "baptisms" caused topographic "anarchy" and inefficiency in fieldwork.[79] Name *changing* was not the only issue. A veritable "surplus of signifiers" saturated a rural landscape already multiply inscribed with the histories of its repeated colonizations.[80] Towns and villages had various names depending upon to whom one talked, which map one viewed, or which set of statistics one consulted: common [*vulgar*] names, indigenous names, ecclesiasti-

76 See, for example, Angel García Peña's report in *Memoria presentada al Congreso de la Unión . . . 1908–1909*, 54.

77 See chapter 1.

78 Díaz, *Memoria de la Comisión Geográfico-Exploradora . . . 1878 a 1879*, 26–27. Díaz was referring to Romero and García Cubas.

79 Ibid.

80 "Surplus of signifiers" comes from Brand, *The Spectator and the City*, 2.

cal names, old or "ancient" names [*nombres antiguos*] as well as varying combinations of these. Existing maps replicated such confusion, and Díaz feared that his own surveyors' maps would depict nothing more than an "entirely imaginary" landscape.[81]

Hopes that local guides could mitigate such confusion were frequently dashed. Twice the Veracruz government felt compelled to issue stern warnings: "[I]t will not be permitted that these [CGE] surveyors will be an object of inconsiderate exploitation on the part of those who should help them as guides or lend whatever assistance they can."[82] In particular, the circulars stressed that guides and individuals be *honest* and *truthful* when showing the engineers "the roads and paths" that cross-hatched the terrain, and that they provide them with "the *true names* of the settlements, ranches, haciendas, hills, rivers, etc."[83] Such reiterations suggest that villagers hired as guides could, and did, easily foil the surveyors' desires, taking advantage of existing ambiguities to mislead them. Although it is difficult to discern with certainty to what degree such actions were conscious evasions, the language of the government's reports clearly suggests that state officials saw them as such.

At the same time, it is worth considering that confusion of this kind did not always indicate resistance but rather was the understandable outcome of a disjuncture between the kinds of questions surveyors asked and the kinds of answers guides could provide. These "ethnographic moments" of ambiguity and confusion frequently arose when surveyors misunderstood the differences between the information they desired and the function of topographic knowledge on the ground.[84] Díaz Ri-

81 Díaz, *Memoria de la Comisión Geográfico-Exploradora . . . 1878 a 1879*, 26–27.

82 Informe de A. Guido, April 8, 1889, CFD, fondo 621, carpeta 1, leg. 0005, doc. 2; and Comisión Geográfico-Exploradora. Circular Num. 6 expedida por el Superior Gobierno del estado previviendo que las autoridades deberán prestar a dicha comisión las guías que fueren necesarios así como ministrar los nombres de los lugares para la formación de los planos, January 29, 1895, AMO, año 1895, caja 214, Fomento. The same circular appears in CLEV—*1895*, 16–19.

83 Comisión Geográfico-Exploradora. Circular Num. 6 . . . January 29, 1895, AMO, año 1895, caja 214, Fomento. My emphasis.

84 On ethnographic moments, see Dening, "A Poetic for Histories," and Michael, "When Soldiers and Statesmen Meet." I am grateful to Bernardo Michael for bringing Dening's article to my attention and for sharing his own work in progress.

vero, on his exploration across the sierra of Huachinango, expressed surprise and suspicion that the guides he had hired were unable to tell him
the precise location of the line dividing the states of Puebla and Veracruz.[85] For officials trying to determine jurisdiction and property rights
at various scales, imprecise boundaries were a fundamental problem intricately tied to the rationalization of fiscal and political rule. But ill-
defined jurisdictional limits were not necessarily a problem for local
populations. Even when limits proved to be an issue of contention, they
could be understood in more ambiguous and contextual, rather than
precise and abstract, terms. The guides' lack of knowledge regarding
state limits may reflect how little importance certain political lines had
for them rather than concerted resistance or willful ignorance. They
may have been as mystified as Díaz Rivero. Similarly, definitive names
may not have been a necessary part of local reality to the degree that
they were for an increasingly "scopic regime." Díaz himself recognized
this. He suggested that the problem may not have been villagers and
guides purposefully confusing the engineers but of guides themselves
confused by "the whims of the Church, the legislator or of history."[86]
The problem, in other words, stemmed as much from the nature of the
activity—of setting down lines and inscribing permanent names—as it
did from the nature of local inhabitants. What appeared of immediate
relevance in the metropolitan cabinets of legislators and governors
faded from view when located in the quotidian reality of village life.

The demands of daily life also led villagers to engage in simple acts of
noncompliance. For example, throughout the 1880s Díaz made repeated
attempts to have pueblos submit geographic information regarding
their location, limits, place-names, and the like, with little success.
Pueblo officials were slow to respond, and their inaction was evidently
tolerated by municipal and cantonal officials, who were berated for responding to such noncompliance with indifference.[87] Without a degree
of state enforcement, or a compelling reason to have such information
compiled and submitted, pueblos simply ignored the requests. Such de-
fiance would not survive in the coming decades, as state governments

85 Díaz Rivero, "Memoria de los trabajos."

86 Díaz, *Memoria de la Comisión Geográfico-Exploradora . . . 1878 a 1879*, 26.

87 Informe de A. Guido, April 8, 1889, CFD, fondo 621, carpeta 1, leg. 0005,
doc. 2.

enforced such orders and as pueblos themselves took the opportunity to stake territorial claims in the midst of increasing concerns regarding village limits and land divisions.

The same held true for orders that municipal authorities provide guides and assistants, as well as housing, to military mappers and CGE engineers. The constant reiteration of such orders clearly reveals that assistance was not provided with regularity.[88] Such noncompliance may have had less to do with the nature of the CGE's work than with the mere fact that housing, feeding, and guiding military surveyors could be an onerous and tension-fraught burden. If the survey crews found themselves in the campo during planting or harvest season, they could hardly count on winning away guides or assistants. The religious, as well as agrarian, calendar also foiled their efforts. When Alvarado and his crew headed into the field in the spring of 1883 to survey a route from Xalapa to Tehuacán, they planned on hiring villagers in places along the route to help with their work. But attempts to enlist help in the town of Perote, west of Xalapa, failed. It was *semana santa* (holy week) and the townspeople were more inclined to look to the heavens for salvation than to help a stranger fix his location.[89]

In sum, the reality of the human and physical landscape complicated attempts to produce stable portraits of it. The contingencies of the ground foiled the quest for a perfectly structured cartographic representation. However, as problematic as the traverse was, it had significant benefits. It was, in contrast to the trigonometric survey, affordable and relatively quick. Moreover, while it may not have achieved the heights of perfection to which Díaz and Porfirian positivists aspired, it did create

88 See ibid. This was a form letter sent out to the municipalities. Two years later, Alejandro Prieto, governor of Tamaulipas, sent to municipal authorities in Tamaulipas a circular that repeated it almost verbatim. See Alejandro Prieto, March 20, 1891, CFD, fondo 621, carpeta 1, leg. 12, doc. 2. These instructions were reiterated a decade later to each canton and municipality. See Circular 5, Sección de Gobernación, April 30, 1901, *CLEV—1901*, 9–10; Comisión Geográfico-Exploradora: Circular número 6, January 29, 1895, AMO, año 1895, caja 214, Fomento. Small municipal councils received the circulars and gave them due consideration. See, for example, "Sesión ordinaria de 22 de Octubre de 1901 del Ayuntamiento de Acultzingo," AMA, libro del Ayuntamiento, 1900, p. 51. It is worth mentioning that I found little evidence of forced labor conscription in the documents.

89 Julio Alvarado to Agustín Díaz, August 15, 1883, ACGE, carpeta 112.

images of the land that far surpassed those available. But perhaps most significant, it incorporated exploration into the process of surveying and map construction. As such, it immersed surveyors in the countryside and permitted them the opportunity not only to survey and map but also to produce a kind of thick description of the regions they traversed.

IV. SITUATED KNOWLEDGE

CGE surveyors did not simply view the lands they surveyed as blank spaces to be inscribed. Mapping and exploration in the national period cannot be reduced to, or equated with, the kinds of forays imperial explorers made into the "blank spaces" of colonial maps and imaginations. CGE surveyors were conscious of the fact that they did not go forth into terra incognita but *terreno vernacular:* places that were part of their own history, punctuated with multiple and competing place-names, layered with a variety of practices, and assigned meaning through both physical labor and the power of imagination. To realize territory as a state required the effective merger of the vernacular, of local knowledge, into the broader complex of understanding. Traverse surveyors were thus charged with more than surveying: they were charged with acquiring and producing *situated knowledge*, a kind of thick description of a region based upon personal observations and experiences but consciously observed through the filters of state desires. This situated knowledge ensured that the mapped image of the land would be more than the uniform and structured location of points upon an undifferentiated terrain; it would be an image linking the grid with the ground and, by extension, the federal state with the rural populations it purported to represent.

The example of naming practices will once again serve as a case in point. A primary task of the surveyors was to permanently put an end to the myriad questions over place-names. Place-names tethered the grid to the ground by providing a unique and differentiated named place to each abstract coordinate. A map with no names would be worthless for administrative purposes. Yet a map with too many names was similarly problematic. Díaz himself remarked that an important dimension of the CGE's work was to "rid the maps of the infinity of words that make them extraordinarily confusing."[90] And not solely the maps. The profusion

90 Díaz, *Exposición internacional colombina de Chicago*, 16.

and confusion of names surveyors confronted in the *field* threatened the translation of landscape to paper, perpetuating the state's incapacity to "see" with confidence. Thus, surveyors were to aid in the determination of the "proper" name of each place, be it a land feature or a population center. Yet Díaz never permitted his surveyors to simply name places of their own accord. Although he felt strongly that the federal government should monitor the changing of place-names, he did not advocate the arbitrary imposition of them. Rather, he charged his surveyors with compiling catalogs of names—of hills, mountains, pueblos, barrios, rivers, and towns—based upon the information provided them by guides, villagers, and documents in archives.[91] These catalogs would then be used by office personnel, who would arrange the collected names into one of five categories: indigenous, legal, common [*vulgar*], religious, or old names no longer in use [*antiguo en desuso*]. Recognized experts in indigenous orthography were then consulted to determine which name constituted the "proper" or "authentic" name of the community or feature in question, to recover "untrackable" names, and to restore "Castilianized" indigenous names to their original.[92]

Indigenous names were almost always given explicit priority on the map.[93] The reasons are fairly obvious. For one, a veritable *indigemanía* pervaded official culture by the 1880s: officials raised a statue of Cuauhtemoc on the Paseo de la Reforma; they devoted an entire tome of the massive *México a través de los siglos* to the preconquest era; and they consciously revived the *X* in place of the *J* in place-names.[94] Indigenous place-names were part of this (admittedly rhetorical) cultural revival

91 See, for example, "Carta general de la República Mexicana: Hoja 19-I-(H). Catálogo de los nombres multiples de poblaciones y lugares que aparecen en la 1a. edición de las hojas publicadas a la 100 000a., conteniendo los indígenas corregidos por el Auxiliar de la Comisión Geográfico-Exploradora, Manuel M. Herrera y Pérez," ACGE, carpeta 112; Informe de A. Guido, April 8, 1889, CFD, fondo 621, carpeta 1, leg. 0005, doc. 2; and Angel García Peña a Don Luis Terrazas, October 27, 1903, ACGE, exp. 4, exp. del Estado de Chihuahua.

92 Peñafiel, *Nomenclatura geográfica de México*, v.

93 Díaz, *Memoria de la Comisión Geográfico-Exploradora . . . 1878 a 1879*, 27.

94 All CGE stationery, for example, spelled Xalapa with the *equis* rather than the *jota*. In 1893 José Miguel Macías (a professor of Greek and Latin in the Colegio Preparatorio del Estado, member of the SMGE and of the Academia Nahuatl) wrote a series of absurd imaginary dialogues about the use of *X* or *J* in the name of the

promoted by the Porfirian elite which served to drape the landscape in an onomastic net of historical association. But indigenous place-names also had practical significance for the administration. Díaz, steeped in the texts of Romero and García Cubas, clearly understood the historical and administrative import of indigenous place-names. These were, after all, the names that had graced centuries of colonial documentation held in the AGN. As conflicts over land rights, boundaries, and jurisdiction proliferated, "original" place-names acquired administrative significance as the means by which to potentially resolve such disputes through references to historical documentation.[95] Moreover, most villagers continued to use some variation of the indigenous name when referring to their villages, although the CGE's experts were not averse to ignoring popular usage when indigenous orthography was at stake. Those names that did not appear on the map, the "suppressed names" (as they came to be known), were preserved in catalogs and forwarded to the Ministry of War. Such names, after all, would not disappear as quickly from the land as they did from the map. They constituted important reference material for military personnel and bureaucrats who would one day need them for field activities.[96]

And here is where the surveyors' situated knowledge assumed powerful proportions. Such local knowledge was key to the expansion of military and political control over the country. The traverse itself was built upon the military traditions of quick reconnaissance, of an incursion into hostile or potentially hostile territory to quickly discern the lay of the land.[97] Traverse surveys, Díaz argued, "would make the [military] very knowledgeable of the territory" and provide it with the kind of visual knowledge considered "indispensable to the art of war."[98] Indeed,

state capital. One of the fictional voices, throwing his weight behind the equis, claimed that "the choice of the Comisión Exploradora [*sic*] is good enough for the rest of the country." See Miguel Macías, *Xalapa o Jalapa?* 55.

95 The decision to keep ecclesiastical names as prefixes to indigenous place-names is telling in this respect. To take one hypothetical: numerous "Acultzingos" may have dotted the countryside, but there was most likely only one *San Juan Bautista* Acultzingo.

96 See Díaz, "Catálogo de los objetos," 16; Alvarado, *Comisión Geográfico-Exploradora . . . París en 1900*, 21–22, 26–27, 40–41.

97 Burnett, *Masters of All They Surveyed*, 86.

98 Díaz, "Informe sobre el estado actual de la cartografía," 477.

until the 1890s CGE surveyors devoted much of their fieldwork to collecting data for the Ministry of War to ensure military capability in fending off foreign invasions and in putting down regional rebellions. Their itinerarios, filled with detailed information on the landscape, became the exclusive property of the Ministry, filed away in a growing topographic archive.[99] They endowed the military with enormous visual power. Eventually, military maps would be produced from the multitude of detailed itinerarios specifically for military operations.[100] In effect, control of the territory derived in part from increasingly exclusive control of geographic information.[101]

CGE surveyors also created pictorial images of, and took copious notes on, the landscape they traversed. They drew views of the countryside to give their itinerarios a visual depth. What an abstract map gained in accuracy with regards to distance, direction, and relative location, it lost in terms of perspective. It might well tell one how to get from Naolinco to Misantla, for example, but one could hardly imagine the density or sparseness of the brush, the location of the best views of the surrounding country, the quickest way to circle a location, or the most dangerous areas of vulnerability. For military personnel who needed to know intimately the lay of the land, a perspectival and horizontal view of the landscape, of the kind they might have when in the field, was critically important. They needed to know what in fact a place looked like (figure 10).[102]

The traverse offered particular military advantages, as surveyors could take extensive notes on the places through which they moved, notes that, along with their itinerarios and images, they forwarded to

99 By mid-1886 Guerra had created sixteen atlases of itinerarios gathered from the CGE. Agustín Díaz to Porfirio Díaz, May 18, 1886, CPD, leg. 11, caja 11, doc. 5036.

100 Díaz, "Catálogo de los objetos," and Alvarado, *Comisión Geográfico-Exploradora . . . París en 1900*, 24–25.

101 See the anonymous letter to the Secretaría de Fomento, March 21, 1893, ACGE, paq. 162. The handwriting, paper, and origin of the letter strongly suggest its author was Agustín Díaz. See also the more general comments in Díaz, "Catálogo de los objetos," and Alvarado, *Comisión Geográfico-Exploradora . . . París en 1900*, 26–27.

102 See Díaz, "Catálogo de los objetos," 22–23. On such imagery and its relationship to survey projects in the U.S. West, see Trachtenberg, *Reading American Photographs*, chap. 3.

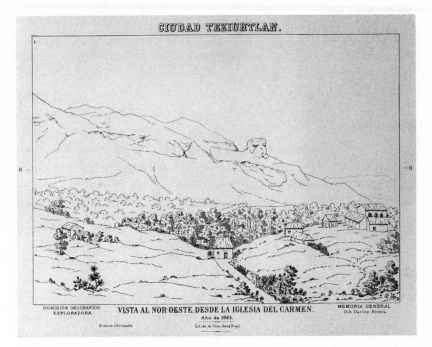

FIGURE 10. The surveyor in place: Anonymous engraving, view of Ciudad Teziuh-tlan, 1885. Courtesy of the Mapoteca Manuel Orozco y Berra, Mexico City. Photograph by Carmen H. Piña.

the Ministry of War.[103] For example, in 1883 Captain Francisco Díaz Rivero conducted an exploration and survey of the region between Xalapa and Boca de Lima (coastal Veracruz), and between Boca de Lima and Huachinango (in the rugged Huasteca).[104] His instructions were comprehensive: to compose a military map of the region, focusing upon the "unexplored places" furthest from the state capital; to acquire an idea of the lay of the land and the practical difficulties of moving through it; and to firmly situate the state boundaries between Veracruz and Puebla in the sierra. His instructions nicely capture the duality of the traverse. On the one hand he was to acquire very practical, experiential knowledge of the ground; on the other, he was to firmly fix a political boundary for inclusion on the map. His resulting narrative provided extensive, detailed information of the landscape to his military superiors.

103 Díaz, "Informe de la Comisión Geográfico-Exploradora," 75; and Agustín Díaz to Porfirio Díaz, May 15, 1886, CPD, leg. 11, caja 11, doc. 5036.
104 Díaz Rivero, "Memoria de los trabajos."

He informed them that the Actopan River, with its calm and deep waters, constituted the easiest means to move artillery and troops between the sierra and the sea, and he constructed a detailed topographic map of the entire river basin over the course of a few days.[105] He discussed the multitude of problems a military detachment could expect to encounter in the region, ranging from a lack of pasture for horses to blowing sand kicked up by the nortes. And he mapped the paths that connected various pueblos in the sierra, where numerous rebels had sought refuge from the government over the years. He was less successful in fixing the state boundary between Puebla and Veracruz. The guides he had hired at, in his opinion, exorbitant costs were unable to provide him with the information necessary to locate the boundary.

Díaz Rivero explored and surveyed a forbidding region, unknown to state officials. However, many of the surveys took place on principal routes in the countryside between relatively medium- or large-sized towns of administrative importance, such as the 1903 traverse through the sierra of Huatusco by an unnamed lieutenant.[106] His narrative similarly stressed issues of military significance: detailed observations of the road he used; population figures of settlements with an eye for the number of able-bodied men who could be pressed into military service; availability of potable water; the size of buildings in the communities, along with estimates as to how many men and horses could be barracked in them; the existence or absence of telegraph lines; and details on the physical landscape itself, such as places where passage would prove difficult during the rainy season, or places along the route where vegetation grew thick enough that it impeded any view of the landscape. His narrative recovered the minutiae of detail excluded from the itinerario and the final map. The existence of potable water, the numbers and kinds of animals in a given household, or the square meters of floor space: these were not what maps were made of. An overabundance of detail impeded the communicative ability of the map. Surveyors had to be selective about what appeared on an itinerary in order to see *both*

105 Ibid.

106 Cuaderno de Tepatlaxco y Huatusco, con varios datos. Itinerario de Paso del Macho a Teptalaxco, May 27, 1903, MOB, CGV, Varilla 18, no. 9759. On surveys in populated regions, see the collection of itinerarios in ACGE, libreta 6, "Itinerarios topográficos de la hoja 19-II-(v) a la 100 000a. Borradores de campo"; and Díaz's comments in "Informe de la Comisión Geográfico-Exploradora," 74.

the proverbial forest *and* its trees. Such details were the kind any military detachment passing through the region would want to have on hand. While a combination of itineraries would, together, form a cartographic space to aid in maneuvers and movements, the narratives and vistas put the military observer in place, with some sense of the experiential rather than the abstract lay of the land. Thus the construction of itinerarios, views of the landscape and narrative records of passage worked in combination to give military officials a comprehensive and fine-grained portrait of the countryside, at least along those lines of travel of intrinsic importance to military exercises. They prepared future military personnel for what lay ahead, oriented them in the field, and provided visual cues for movement in an unknown landscape.

Herein lay the true value of the traverse survey: in the indispensable and by all accounts irreplaceable situated knowledge that military surveyors, fiercely loyal to don Porfirio, acquired of the physical and social topography. Don Porfirio and the state governor, Teodoro Dehesa, both understood that the static map could never accurately capture the landscape to the same degree as the fieldwork that produced it. They knew perfectly well that maps were too rigid to be real. In this sense, the most valuable knowledge that men such as Canóvas, Díaz Rivero, and Huerta held for don Porfirio's administration was that which had *not* survived translation from the field to the map: the situated knowledge garnered of a region based upon the experience accrued from moving through it.

Such knowledge made them excellent candidates for politically powerful positions, such as prefectures, in the regions where they worked.[107] In 1899 Rafael Rosas, Ignacio Muñoz's close friend from the Colegio Militar, became prefect of the canton of Papantla. He had arrived there in 1895 as a member of Muñoz's CGE crew, charged with completing the land division and quelling the significant unrest that pervaded the region. In that same crew were Francisco Canóvas and Alberto González. González soon after assumed the powerful position of prefect in Xalapa canton.[108] Canóvas had been appointed to the CGE in 1892 and promptly

107 On the prefecture (or jefatura política), see Falcón, "Force and the Search for Consent"; and Knight, *The Mexican Revolution*, 1:25–30.

108 On González's participation in the Papantla surveys, see José de la Luz y Soto to Gobernador del Estado, July 3, 1895; Instrucciones que se dan al Ingeniero C. Ignacio Muñoz, August 19, 1895; and Ing. Alberto González to Jefe Político

FIGURE 11. Situated knowledge and political power:
Francisco Canóvas y Pasquel, CGE surveyor and prefect of Misantla, 1905. Courtesy of Carmen Boone
de Aguilar and Daniela Canovas Rebling.

been sent to Veracruz, where he worked for a decade, primarily in regions around Papantla and Misantla.[109] Upon departing the agency in
April 1902, he assumed the post of prefect of Misantla, a position he held
until the revolution forced him from power in 1911 (figure 11).

Clearly more than their work in the CGE determined their appointments to such powerful positions. Canóvas, for example, was a member

Angel Lucido Cambas, October 24, 1895; all in AGEV, RGYJ, Tierras: Comisión Ingeniero, División de Terrenos, caja 11, general 2414, 1895–1905.

109 On Canóvas's appointment to the CGE, see Secretaría de Guerra y Marina,
November 18, 1892, FCP. His departure date comes from Juan B. Laurencio to
Francisco Canóvas, April 9, 1902, FCP. On his last surveys in the CGE, see "Carta de
la República Mexicana . . . 1a Serie; hoja 19-II-(H). Terminada en 1902, publicada
en 1903," SMGE Map Collection, no. 3286. The relationships these individuals
forged with one another persisted beyond their time in the CGE. When Rosas became prefect in Papantla, Canóvas became the president of the Junta de Sanidad.
See Figueroa Domenech, *Guía general descriptiva de la República Mexicana*. Under
Canóvas's tenure, Muñoz surveyed communal lands in Misantla canton. See Francisco Canóvas to Secretaría del Gobierno, July 7, 1902, AGEV, CLA, caja 1, exp. Yecuatla, ff. 172r–74v and ff. 223r–24v.

of a prominent Xalapan family. His father was Dr. Sebastián Canóvas y Pérez de Tudela, founder of the Hospital de Caridad (Xalapa's present-day Hospital Civil), vice consul of Spain, and regidor of the capital's ayuntamiento. His mother was descended from two politically and economically important families from Veracruz's Sotavento region.[110] In 1898 he married Esther Bouchez Echeagaray, daughter of two very prominent xalapeño families (figure 12). Such familial networks and political connections were fundamental considerations in the distribution of politically powerful positions. Yet it is hard to ignore the fact that Canóvas, as well as other CGE surveyors, assumed positions as prefects in the very places where they had conducted many, if not most, of their surveys. Who better to appoint as a regional intermediary between, on the one hand, federal and state authorities and, on the other, village and municipal authorities, than an individual loyal to, and politically invested in, the regime and at the same time intensely familiar with the physical, social, and political lay of the land?

CONCLUSION

In 1896 Francisco Díaz Rivero launched a systematic critique of the CGE's work.[111] The brunt of his criticism fell upon the method adopted in 1877 by his uncle, Agustín Díaz, to map the country. To wit: that the instruments and method used were fine for land exploration but not suitable for the creation of a carta general; that traverse surveys were not an efficient means for comprehensively mapping a country the size of Mexico; that local maps of dubious quality were used to fill in the empty cartographic spaces; and that the astronomical method of determining points on the ground was very imprecise.[112] Díaz Rivero concluded with a blunt technical invalidation of the CGE's entire corpus:

> [T]he number of points that constitute the *canevas* [geographic grid] is insufficient. They [the points] have no geometric relationship to one another. They have not been chosen in conspicuous places with conditions of good visibility. They do not link the boundaries of

110 Carmen Boone de Aguilar, "Cronología de Francisco Canóvas y Pasquel (1870–1954)," manuscript in author's possession.

111 Díaz Rivero, *Estudio preliminar*.

112 Ibid., 14–19.

FIGURE 12. Marriage of Francisco Canóvas y Pasquel and Esther Bouchez y Echeagaray, January 30, 1898. Courtesy of Carmen Boone de
Aguilar and Daniela Canovas Rebling.

private property.... The manual operations of constructing and drawing the maps are laborious and extremely slow, [and the resulting images] do not reflect the reality of the land due to the irregularity of the field work. The personnel of the Commission is not properly trained and recruited. [In sum,] the maps of the Commission cannot fulfill even slightly the demands and services to which they are called.[113]

As devastating a critique as this was, Díaz Rivero's exegesis missed the point. His fetish for technical proficiency obscured from view the political exigencies of early Porfirian Mexico. For all the rhetoric, when he supported the formation of the CGE, don Porfirio was not solely concerned with a technically perfect map of the Republic. The issues of most pressing urgency were those of political stability, state centralization, and administrative control. Moreover, state officials were perhaps less enamored with the "panoptic" ideal than one might think. They understood that the panoptic possibility of the state map was of limited potential if removed from the experiences and itineraries that were the "condition of its possibility."[114] Improving geographic knowledge meant not only better [*más fiel*] images of the country acquired through surveys but also more comprehensive reports about the countryside itself through exploration. It meant not dismissing local knowledge but validating and incorporating it. It meant the creation of an archive of structured geographical knowledge, premised upon improved methods of data collection, surveying, and mapping, but also the formation of a cadre of officials with the situated knowledge needed to rule more effectively.

In the final analysis, the inappropriateness in 1896 of mapping strategies devised in 1877 derived less from technical or financial issues than from the fact that don Porfirio's Mexico had changed. By the late 1890s the old concerns of military knowledge, regional geographic information, and social control had been superseded by new ones: infrastructural development, property registration, and a growing agrarian question. This shift is the subject of the next chapter.

113 Ibid., 48–49.
114 de Certeau, *The Practice of Everyday Life*, 120.

CHAPTER 5

Spatial Progressions:
The Geographic Exploration Commission (II)

In 1884 the Porfirian statesman Alfonso Lancaster Jones composed an introduction to Antonio García Cubas's *Cuadro geográfico, estadístico, descriptivo e histórico de los Estados Unidos Mexicanos*, a text destined for the New Orleans Exposition and intended to promote Mexico as a place for foreign investment and colonization.[1] In his introduction, he did more than extol the virtues of García Cubas's tract and sing the praises of don Porfirio; he offered a short history of Mexico, from pre-Cortesian times to the present, scripted as the nation's progression from infancy to maturity, or, in the lexicon of the day, from barbarism to civilization. In a remarkable passage, he dismissed the violence of the conquest as a regrettable but necessary episode in the history of civilization. "The painful path of progress," he wrote, "revealed itself to America's primitive nations through the bloodied steel of the Cortés and the Pizarros."[2] Drawing liberally from Goethe's *Faust*, he turned historical act into philosophical principle, rescuing civilization from the stain of its own barbarism: "I am part of that force which forever wishes evil, but which always and in spite of itself does good."[3]

Four hundred years after the arrival of Cortés and Pizarro, conquests continued, this time under the Porfirian maxim of Order and Progress, a schizophrenic idealization of both fixity and flux around which ideas

1 García Cubas, *Cuadro geográfico, estadístico, descriptivo, e histórico.*

2 Ibid., xii.

3 Ibid., xi. The lines (part 1, lines 1335–36) are spoken by Mephistopheles as an answer to Faust's question: "Who are you, then?" The Spanish original reads: "Yo soy aquella parte de la fuerza que siempre quiere el mal, pero que siempre y a su pesar hace el bien."

of advancement orbited. Darwinian emphases on evolution and process were accompanied by a concomitant concern with pattern and order. Order meant progress; progress required more order. In this chapter, I look at the imposition, codification, and display of spatial order as manifested in the various works of the Comisión Geográfico-Exploradora in the final decades of its existence. In part 1 I examine two CGE military operations designed to clear the path of progress of its remaining "primitive" detritus. Specifically, I look at the militarization of the land division process in the state of Sonora and in the region of Papantla, Veracruz. In part 2 I turn my attention to the role of the CGE in the continuing quest for spatiojuridical order by analyzing its role in the attempts to fix municipal, cantonal, and state limits and to "discover" places, both of which would spur agricultural development, foreign investment, colonization, and commercial production for the international market. In part 3 I examine the final maps of the CGE to suggest how iconographic and textual codes—the politics of the image—helped create a certain image of politics and stability. That Lancaster Jones wrote an introduction for a text penned by a cartographer and destined for an international exhibition is fitting. The images circulating on the world's fairs, including those of the CGE, fashioned Mexico as a place of political, social, institutional, and legislative stability, as a world amenable to economic rather than territorial conquest.[4]

I. DIVISION

In May 1886 Agustín Díaz sent a personal letter to don Porfirio, proudly reporting the CGE's successes in the field.[5] They had prepared for publication eighteen sheets of the carta general; another eight, as well as a large number of urban maps, were under construction; and sixteen atlases of itinerarios had been submitted to the Ministry of War's topographic archive. After devoting some words to the CGE museum recently opened in Tacubaya, Díaz concluded by requesting permission to present the president's wife with a small gift in the name of the Commission.

In response, don Porfirio offered a terse congratulation and accepted the gift. He then took "advantage of the moment to raise a thought I

4 See Riguzzi, "México próspero," 137–58.
5 Agustín Díaz to Porfirio Díaz, May 18, 1886, CPD, leg. 11, caja 11, doc. 5036.

have had": to deploy a commission of surveyors to divide the communal lands of the Yaqui and Mayo Indians in Sonora.[6] The work, he wrote, required a staff "above all suspicion with regards to their loyalty and who would carry out faithfully the instructions that I will personally give them."[7] Who, he asked Díaz, would fit the bill? Soon after, Díaz and a number of his CGE staff prepared for their departure. Financial shortfalls and a concerted resistance by the head of the Sonoran military campaigns against the Yaqui, General Angel Martínez, who considered any attempt to divide lands as premature, delayed the expedition's arrival. But after a series of personal interventions by don Porfirio, the Sonoran Scientific Commission finally arrived at its destination in June 1887.[8]

Díaz's first act upon arriving in the Yaqui valley was to deliver a speech to an undisclosed number of Indians:

> I come in the name of the president of the Republic, General Porfirio Díaz, to reorganize your pueblos and to secure your happiness. In his name also I offer you all classes of guarantees. Return without fear to your homes that you have abandoned and peacefully dedicate yourselves to working for your families. . . . The president said to me these words when I left his side to be among you: "Go to the Yaqui and Mayo rivers, and assure the Indians of their full right to the lands that you will mark for them, to cultivate them and construct houses with the comforts that civilization requires. . . . Direct them in their labors such that they might enjoy the benefits of progress and bring them to civilization such that they might feel the well-being it produces. . . . Treat all the Indians as our brothers, and once you know their problems, propose to me the means you believe adequate to remedy them."[9]

6 Porfirio Díaz to Agustín Díaz, May 21, 1886, CPD, leg. 11, caja 11, doc. 5037.
7 Ibid.
8 The Treasury severely limited the agency's operating budget at the outset and then cut its funding, leaving the entire expedition stranded in El Paso. By that time the staff had not been paid in months. Agustín Díaz to Porfirio Díaz, April 30, 1887, CPD, leg. 12, caja 6, doc. 2867; Agustín Díaz to Porfirio Díaz, May 3, 1887, CPD, leg. 12, caja 8, doc. 3817; Agustín Díaz to Porfirio Díaz, May 25, 1887, CPD, leg. 12, caja 8, doc. 3818; and Angel Martínez to Porfirio Díaz, May 27, 1886, CPD, leg. 11, caja 13, docs. 6084–85.
9 Agustín Díaz, Coronel de Estado Mayor Especial y Jefe de la Comisión Científica de Sonora, to los indígenas de los ríos Yaqui y Mayo, June 10, 1887, CPD, leg.

The remedy had already been prescribed. Díaz's speech alluded to the tasks don Porfirio had charged him with performing: organize the Indians into settled pueblos with a traditional plaza and grid pattern and then divide the lands into individual lots for both Indians and potential colonists, who would receive land and formal titles at no charge.[10] Don Porfirio viewed both operations—forced settlement and the land division—as essential components to the pacification and civilizing of the Yaqui.[11] They were spatial fixes to a persistent Yaqui resistance to Porfirian order and progress. Social order and capitalist progress rested on the imposition of a new spatial order with its own moral, political, and economic modalities. No longer would Yaqui and Mayo Indians be permitted to "live, according to their customs . . . independently, choosing as they wished the lands that pleased them most," an offense to the meritocratic facade of liberal thinking.[12] Rather, they would be "subjected to a determined area," surveyed, and fixed.[13] Whether or not the Yaqui and Mayo had always lived "independently" is open to question, but for men such as Martínez it appeared unthinkable that the semisedentarist lifestyle of the Yaqui and Mayo could itself have been a response to the disruptions introduced by capitalist state formation. Like Lancaster Jones, Martínez assumed a philosophy of history and geography that conceived of non- or semisedentary lifeways as a primitive antecedent to, and antithesis of, modernity rather than a response to its destabilizing power.[14] Their topology had a chronology: only in being "settled"

13, caja 3, doc. 1497. In 1908 Francisco de P. Piña, in a speech praising Agustín Díaz, would suggest that Díaz gave the speech in "an Indian dialect." I have found no evidence of this anywhere. See de P. Piña, "La Comisión Geográfico-Exploradora y la influencia de sus trabajos."

10 Agustín Díaz . . . to los indígenas de los ríos Yaqui y Mayo, June 10, 1887, CPD, leg. 13, caja 3, doc. 1497; Hu-DeHart, *Yaqui Resistance and Survival*, 120. See also Hernández Silva, *Insurgencia y autonomía*, and Dabdoub, *Historia de El Valle del Yaqui*.

11 Porfirio Díaz to General Angel Martínez, May 3, 1887, CPD, leg. 12, caja 9, doc. 4084; Porfirio Díaz to José F. Otero, January 13, 1887, CPD, leg. 12, caja 1, doc. 26; and Otero to Porfirio Díaz, January 29, 1887, CPD, leg. 12, caja 2, doc. 854.

12 Angel Martínez to Porfirio Díaz, May 27, 1886, CPD, leg. 11, caja 13, docs. 6084–85.

13 Ibid.

14 For a careful examination of such tropes, see Fabian, *Time and the Other*, and O'Brien and Roseberry, eds., *Golden Ages, Dark Ages*.

would the Yaqui settle down, cultivating the land and, by extension, their own evolution.

Spatial subjugation would be rigorously rational. Each new settlement would have a central plaza with rectilinear streets running off it; each land parcel would be geometric and uniform. Neither ideal was particularly new: colonial towns were plotted out as grids arranged around a central plaza, whereby proximity to the plaza indicated political power. By the nineteenth century, rational grid plans were mirrors of modernity itself, an ideal captured equally in Díaz's mammoth plan for Mérida and in Holzinger's reconstruction of Acultzingo's fundo legal such that it would have "the symmetry which Modern Towns require."[15] In the case of Sonora, don Porfirio himself weighed in on what a "good lot" would look like: a rectangle, with a length fifteen times the width, running from the village to the river. Each lot granted to an Indian family would be located between lots put aside for "civilized people"—or "colonizers"—who would help "spur the Indians' moral advancement."[16] Díaz modified don Porfirio's recommendations only slightly, surveying out square lots of three hectares each for cultivation [*sembradura*] and substantially smaller rectangular urban plots [*solares*].[17] Land recipients would receive plots based on the size of their family and civil status.[18] Ever concerned with fiscal clarity, Díaz organized the lots into a structure of larger quadrilaterals according to an intricate numbering system for easy cadastral and numerical consultation, not unlike the township grids laid out in the United States by the federal Land Ordinance of 1785.[19] Finally, he recorded the location of boundary markers according to their geographic coordinates.

15 On Díaz's Mérida project, see chapter 4n11; on Holzinger's survey and comments, see chapter 3.

16 Porfirio Díaz to Angel Martínez, May 6, 1886, CPD, leg. 11, caja 10, doc. 4595–96. To attract colonists, the state government would provide them with tools and seed at no cost and pay them to settle in pacified regions. Ramón Corral to Porfirio Díaz, February 15, 1888, CPD, leg. 13, caja 4, doc. 1699.

17 Díaz, Distribución de terrenos, en las Colonias de los Ríos Yaqui y Mayo, ACGE, carpeta 106, Cartera con datos y tablas para Memoria de las Colonias de los Ríos Yaqui y Mayo, Son.

18 Ibid.

19 See Díaz, Registro de distribución de los lotes de sembradura en la Colonia de Torin, in ibid. To help standardize the entire process and avoid future ambiguities, Díaz undertook a lengthy examination and study of the relationship between

Díaz's careful designs did not have the desired effect. Angel Martínez, enraged at Díaz's public comments regarding Yaqui participation in the French Intervention, wrote a furious letter to don Porfirio notifying him that Díaz had sat in Buenavista for six months and had seen no more than fifty Indians.[20] By 1890, however, the process picked up momentum under the aegis of Díaz's subordinate, Angel García Peña.[21] It accelerated in 1891, when private companies took over the survey, sale, and colonization of lands. "Civilization" arrived in the form of the U.S.-controlled Sonora and Sinaloa Irrigation Company, partnered with Sonoran businessman Carlos Conant, who, with credit from García Peña, soon controlled some ninety thousand hectares of land in the valley.[22]

In early 1892, five years after the Sonoran expedition, a team of surveyors culled from the CGE and under the leadership of Victoriano Huerta arrived in the canton of Papantla, Veracruz. Governor Juan de la Luz Enríquez, faced with increasingly violent confrontations over the land division there, had requested that the Ministry of War send a crew of military surveyors to carry out the land division.[23] Upon arrival,

---

colonial, vernacular, and modern measurements. Díaz, "Relación entre las medidas agrarias antiguas y las modernas," in ibid. The U.S. system garnered increasing attention in Mexico. See, for example, Gómez, "Concursos científicos."

20 Angel Martínez to Porfirio Díaz, January 27, 1888, CPD, leg. 13, caja 3, docs. 1494–96. Díaz, in his speech, had suggested that "you can be certain that [we] will treat you with the fraternity that is natural between sons of the same race, because we know that your ancestors were our ancestors; nor can we forget the services that along with us you have given to the country, when the foreign intervention wanted to humiliate us." Martínez, a military general in Sonora at the time of the Intervention, personally attested to the fact that the Yaqui had fought *for* the French. He also chafed at Díaz's treatment of him. Díaz had refused to consult him and thought him uneducated and stupid: "Although it is true that I have never entertained the pretension of passing myself off as a wise man, it is also true that with regard to the affairs of the 'Yaqui' and 'Mayo' I believe I have very sufficient knowledge, which experience has given me, to be able to deal with them with a certain amount of insight, and which I would happily put at Sr. Colonel Díaz's disposition if he wanted it." Ibid.

21 Hu-DeHart, *Yaqui Resistance and Survival*, 127–28.

22 Ibid., 128, 161; Hernández Silva, *Insurgencia y autonomía*, 130; Dabdoub, *Historia de El Valle del Yaqui*, 259–93.

23 On the Papantla land divisions, see Kourí, "The Business of the Land," esp. chap. 6; Chenaut, *Aquellos que vuelan*; Chenaut, "Fin de siglo en la costa totonaca"; and Velasco Toro, "Indigenismo y rebelión totonaca de Papantla."

Huerta and his staff (which included the president's nephew Félix Díaz), signed a series of contracts with a number of the *juntas directivas* (administrative councils) of the *grandes lotes,* who were to cover the costs of the surveys.[24] Work progressed slowly, and when don Porfirio recalled the crew in October 1893, they had surveyed only four of Papantla's seventeen grandes lotes. Two years later, this time at the behest of Governor Teodoro Dehesa, a new team of CGE surveyors journeyed to Papantla to divide the lands, a team led by another of don Porfirio's nephews, Ignacio Muñoz, and including Rafael Rosas and Francisco Canóvas. (Canóvas is depicted surveying in Papantla in figure 13.) Again contracts were signed with the juntas directivas, who had been strongly urged, if not required, to contract *only* with the military surveyors. The administration feared that other contracts were merely a means to delay the survey or that civilian surveyors could not be trusted.[25] Some welcomed the surveyors: heads of Zapotal y Carzonera renamed their congregación Colonia Ignacio Muñoz![26] But in other instances their operations provoked further strife. In June 1896 Totonac Indians from a number of grandes lotes descended upon the town of Papantla, at that time empty of federal troops, who were out in the fields protecting the survey crews.[27] The revolts were soon quelled, the land divisions continued, and, not unlike the case of Sonora, by the century's end a number of powerful individuals possessed hectares in the thousands.

The land divisions in Sonora and Papantla, each in their own way, represented something new in Porfirian Mexico: federal military intervention in the process of land division. In the 1880s CGE surveyors had contracted with municipal authorities to divide village lands. Victoriano Huerta, for example, combined land division surveys with his

24 Kourí, "The Business of the Land," chap. 6; Depto. de Guerra y Marina to Gobernador del Estado de Veracruz, January 25, 1892, AGEV, RGYJ, caja 4, general 2408, 1888–1898; Julio Alvarado to Félix Díaz, January 29, 1892, CFD, fondo 621, carpeta 1, leg. 19, doc. 1.

25 See for example, Angel Lucido Cambas, Jefe Político de Papantla, to Secretario de Gobierno, October 24, 1895, exp. Ingenieros. See also the reply in Telegrama de Teodoro Dehesa to Jefe Político de Papantla, October 25, 1895, in exp. Ingenieros.

26 March 4, 1897, AGEV, RG, Tierras, Reparto, exp. Gutiérrez Zamora, 1902, exp. Terrenos. Lo relativo al reparto individual del lote no. 5 denominado "Zapotal y Carzonera."

27 Kourí, "The Business of the Land," chap. 6.

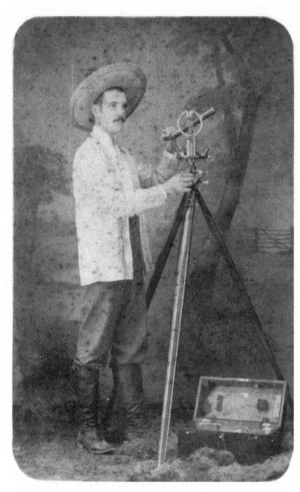

FIGURE 13. Dividing lands: CGE surveyor Francisco Canóvas y Pasquel surveying in Papantla, May 1896. Courtesy of Carmen Boone de Aguilar and Daniela Canovas Rebling.

traverses through Misantla. But these were, as far as can be ascertained, largely individual initiatives rather than state-directed orders. The surveyors drew up personal contracts between themselves and the village authorities; they, not the CGE, received payment for services rendered. The land surveys in Sonora and Papantla, as well as others that succeeded them, were a different matter all together. These came under federal military auspices, carried out exclusively by a crew of military personnel, drawn almost entirely from the ranks of the CGE, at the behest of some of the most powerful political figures of the time.[28] In Papantla the work of CGE surveyors unfolded under the protective eyes of their military colleagues and battalions deployed from the port of Veracruz. As massive firepower quelled the rebellions, Muñoz, Rafael Rosas, and Alberto González hastily completed the land divisions, some in just one day.[29] Meanwhile, the Scientific Commission of Sonora, notwithstanding its euphemistic name, constituted a full-fledged military expedition. Despite the bitter animosity between Díaz and Martínez, each personified a different dimension of a unified and violent process of pacification. "Restructuring" pueblos and "civilizing" Indians through forced settlement constituted a particular form of violence, even if achieved with a surveyor's instrument rather than a soldier's gun.

II. DEVELOPMENT

The CGE's presence in Sonora and Papantla represented an increasing federal involvement in, and militarization of, the land division process.

28 Neither Díaz nor Alvarado saw the interventions in Sonora and Papantla as military activities divorced from the tasks of the CGE. The personnel who served were viewed *as* CGE staff, operating in their capacity *as* CGE surveyors. Alvarado, *Comisión Geográfico-Exploradora . . . París en 1900*, 15, 37; Alvarado, "Informe de la Comisión Geográfico-Exploradora," 299. For the personnel roster, see Depto. de Guerra y Marina to Gobernador del Estado de Veracruz, January 25, 1892, AGEV, RGYJ, caja 4, general 2408, 1888–1898. For another example of how CGE personnel were increasingly used to resolve problematic divisions, see the case of Juan B. Ulloa in Ayahualulco, AGEV, RG, Tierras, Límites, 1902, exp. Lo relativo al arreglo de la cuestión pendiente entre los Municipios de Yxhuacan y Ayahualulco. Military personnel were ordered in to protect Ulloa during his survey. See Teniente Coronel Miguel Girón to Gobernador del Estado de Veracruz, April 11, 1900, AGEV, RG, Tierras, Reparto, 1893–1902, exp. Terrenos: Lo relativo al reparto de los de Municipio de Ayahualulco.

29 Kourí, "The Business of the Land," chap. 6. For details on personnel, see

It also signified a gradual expansion of the CGE's ambit during the last decade of the century. Exploration and traverse surveys were now complemented by a growing emphasis on development and property surveying. This trend is revealingly captured in the changing bureaucratic location of the agency. In 1880 the Ministry of War was the agency's primary patron; by 1910 it had become a dependency of the Ministry of Development's Agrarian Office.[30] Two factors account for this shift: increased attention to property issues on the part of federal officials and the role of state administrations, with their own objectives, in financing the CGE.

Minister of development Manuel Fernández Leal saw great potential in the CGE. For him, it was crucial to the "development of agriculture and industry, to attracting capital . . . [and] to encourag[ing] domestic and foreign [investors] to exploit our resources."[31] Moreover, with the aid of the CGE, a property cadaster, a *Gran registro de la propiedad*, finally appeared as a truly realizable goal.[32] It promised the possibility of determining the limits of property "in a fixed and irrevocable way" and of resolving a "great variety of questions related to private property and that of the State."[33] CGE officials supported Fernández Leal's claims. Díaz Rivero remarked that the CGE had thus far not fulfilled "certain public services *that it should satisfy*," such as producing cadastral maps.[34] Alvarado, now heading the agency after Agustín Díaz's death in 1893, was sharply attuned to such changes and emphasized in 1899 that the agency's earlier operations laid a firm foundation for the Republic's future by facilitating cadastral surveys.[35] In fact, the CGE's Papantla surveys served as the basis for a cadastral map of the entire canton, and one of the CGE's surveyors in Papantla claimed that his map would offer "the

José de la Luz y Soto to Co. Gobernador, July 3, 1895, exp. Ingenieros; and Instrucciones que se dan al Ingeniero C. Ignacio Muñoz, August 19, 1895, exp. Ingenieros.

30 García Martínez, "La Comisión Geográfico-Exploradora," 542. On the growing concern with agricultural development and modernization, see Zuleta, "La invención de una agricultura próspera."

31 García Martínez, "La Comisión Geográfico-Exploradora," 541.

32 *Memoria presentada al Congreso de la Unión . . . 1892 a 1896*, 125–26.

33 *Memoria presentada al Congreso de la Unión . . . 1897 a 1900*, 115–16.

34 Díaz Rivero, *Estudio preliminar*. My emphasis.

35 Alvarado, *Comisión Geográfico-Exploradora . . . París en 1900*, 9–14.

merchant and businessman . . . information that no 'Reporter' could give him."[36]

It was not only federal officials who saw such promise. State administrations stood to gain by having their respective states mapped. Already in 1890 the CGE had attracted the attention of the governor of San Luis Potosí who, that same year, became the first governor to contract with the Ministry of Development to have the CGE create a map of his state.[37] The Ministry of Development favored such an arrangement as it potentially promised both the completion of the carta general and the production of maps of individual states.[38] By the end of 1894 the CGE had contracted with both Nuevo León and Veracruz; Tamaulipas followed soon after.[39] The decision to go ahead with state contracts proved propitious, as the state governments sustained the CGE when, with the economic crises of the 1890s, it lost much of the federal funding and personnel it had laboriously amassed over the previous decade.[40] State governments financed the CGE to the tune of 500 pesos per month during the course of field operations and 100 pesos per month once fieldwork had been completed and only office work remained.[41] Veracruz paid slightly more

36 Manuel Alvarado to Ignacio Muñoz, June 1, 1900, exp. Ingenieros. "Reporter" is an allusion to the *Oil, Paint, and Drug Reporter*, a major business publication at the time.

37 Alvarado, *Comisión Geográfico-Exploradora . . . París en 1900*, 5.

38 The suggestions here come from Angel García Peña to don Luis Terrazas, Gobernador del Estado de Chihuahua, July 16, 1903, ACGE, exp. 4, exp. del Estado de Chihuahua.

39 Alvarado, *Anexo de Memoria*, 223; Alvarado, *Comisión Geográfico-Exploradora . . . París en 1900*, 5.

40 The Memorias of Hacienda, which previously had listed amounts for different agencies, had by 1890 stopped such practices, in large part due to the vast increase in statistical information available. Evidence for the cuts in funding are provided by the statements of various state officials. See, for example, Alvarado, *Comisión Geográfico-Exploradora . . . París en 1900*, 4–7; *Memoria presentada al Congreso de la Unión . . . 1892 a 1896*, 30. See also García Martínez, "La Comisión Geográfico-Exploradora," 509–10. On the composition of the staff prior to the cuts, see Díaz, "Catálogo de los objetos," 6; Alvarado, "Catálogo de los objetos que componen el contigente de la expresada comisión, precedido de una reseña abreviada sobre su organización y trabajos. Exposición del Congreso Geográfico Internacional de Londrés," ACGE, paq. 162.

41 García Peña to don Luis Terrazas, Gobernador del Estado de Chihuahua, July 16, 1903, ACGE, exp. 4, exp. del Estado de Chihuahua.

—800 pesos per month, or 1.5 percent of the state's total budget in 1899 —due to the unhealthy climate of the coast and the exceptional number of rivers to be mapped.[42] While it may not have amounted to much in terms of the state's budget, to the CGE the amounts were critical. Over the course of nine years, Veracruz alone provided the CGE with more than 62,000 pesos.[43]

State governments pushed the project forward in other ways. The Ministry of War cut military escorts for CGE personnel in 1893, making surveying activities both more dangerous and arduous. Yet civilian escorts proved unreliable, as Alvarado vehemently noted in 1894: they quit as soon as they found out they would be working on the coasts of Veracruz or ran off at the first opportunity once they had been paid.[44] Even worse, in cases of aggression, they felt no obligation to come to the surveyors' defense. They were not willing, he grumbled, "to sacrifice their life in defense of national interests," especially for an undertaking "whose utility they did not understand" (although one might speculate that in fact they understood it all too well).[45] In sharp contrast, he wrote, the disciplined soldier understood that any action considered "insubordination, aggression against a superior, or robbing the interests of the Nation" would cost him his life.[46] As a result, the state government agreed to subsidize the cost of twelve escorts, although it is unclear as to whether or not they were culled from the military.[47]

The potential benefits of the CGE's operations clearly outweighed the expense for state administrations. As well as constituting only a small fraction of the state's budget, costs could be recuperated through the sales of the resulting maps, as Angel García Peña, by then director of the CGE, informed the governor of the state of Chihuahua, Luis Terrazas, in 1903.[48] Attempting to convince Terrazas to sign a contract with the CGE,

42 Ibid.; and Ley 38, December 22, 1898, CLEV—*1898*, 170.

43 Angel García Peña to don Luis Terrazas, October 27, 1903, ACGE, exp. 4, exp. de Chihuahua.

44 Julio Alvarado to don Manuel Fernández Leal, July 2, 1894, ACGE, exp. 4, exp. de Veracruz.

45 Ibid.

46 Ibid.

47 Angel García Peña to don Luis Terrazas, July 16, 1903, ACGE, exp. 4, exp. de Chihuahua.

48 Ibid.

García Peña held up San Luis Potosí as an example: the first edition of that state's map had already sold out.[49] He also emphasized how the agency could concentrate its efforts on mapping areas of industrial importance, evaluating property, and surveying rivers with hydraulic energy potential.[50] Such possibilities attracted state governors, themselves not immune from the financial crisis. Dehesa, affirming the value of CGE operations in Papantla, observed that the financial crisis required that the state "stimulate, with all its power, the development of agriculture and industry, as the best and most efficient means to stave off dangers of a similar kind in the future."[51] Development had become the order of the day.

So too had "discovery." CGE personnel, under state government auspices, devoted their energies to "finding" population settlements and property missing from the state's statistical database. With the CGE now in its employ, the government of Veracruz in 1895 ordered every municipio to submit information on all populated points within its confines. Every settlement, "even if there be only an isolated family," had to be assigned a political category (such as *congregación, ranchería, hacienda*) and a name. "Any ranchos and isolated points that have no geographic names," the circular stated, were to be "designate[d] by the name of the inhabitants."[52] The old commonplace ascriptions of *sin nombre* (nameless) and *sin categoría* (without political designation) were now unacceptable. To refer to a place in such terms constituted it as *no place*, beyond the official field of vision. In contrast, names and categories brought such places into being. More important, it brought them into being as objects of state knowledge—as political categories. Increasingly, CGE reconnaissance and mapping expeditions spent their time comparing what they found on the ground with what appeared in state statistical files.[53] Between 1900 and 1904 CGE personnel circulated charts to cantonal and municipal authorities to double-check the names and places

49 Ibid.
50 Ibid.
51 Gobernador del Estado to Secretario de Estado y del Despacho de Guerra y Marina, July 4, 1895, exp. Ingenieros.
52 Circular 9, February 15, 1895, CLEV—*1895*, 29–30. See also Circular 22, October 27, 1899, CLEV—*1899*, 52.
53 See *Memoria presentada al Congreso de la Unión, 1907–08*, 67; *Memoria presentada al Congreso de la Unión, 1908–09*, 57.

included therein.[54] By 1908 Angel García Peña proudly reported that his staff had "discovered" some 2,841 "extra places" in the state of Veracruz alone, over one-third more than existed in the state's own statistics (7,278).[55]

As well as plotting points, state governments sought to fix lines in the hopes of resolving persistent questions regarding political jurisdiction and rural property. Once again, now that the CGE worked for his administration, the governor of Veracruz issued orders that municipal authorities provide CGE personnel with the information, instruments, and access to official buildings necessary to their work as well as all information related to cantonal limits, which CGE surveyors were explicitly charged with mapping.[56] A second decree followed soon after, ordering municipal authorities to furnish detailed information on, and legal evidence of, their boundaries to the agency.[57] This was not the first time such orders had been given. Similar requests had been made in the past to no avail. Díaz had complained as early as 1878 that landowners, villagers, and municipal authorities resisted providing "even the most

54 Poblados que faltan en las hojas, 1903, ACGE, exp. 7; Estadística de varios cantones del E. de Veracruz, October 21, 1902, ACGE, exp. 3; "Sesión ordinaria," AMA, libro de Ayuntamiento, 1900, ff. 76–77; and Actas de sesiones del H. Ayuntamiento, comienzan en 4 de enero y terminan en 27 de diciembre, 1902, AMM, caja 1902, ff. 137, 143–45.

55 Response of García Peña to García Granados, "Discurso pronunciado por el Socio Ingeniero Ricardo García Granados," 312.

56 Comisión Geográfico-Exploradora Circular Num. 6 expedida por el Superior Gobierno del estado previviendo que las autoridades deberán prestar a dicha comisión las guías que fueren necesarios así como ministrar los nombres de los lugares para la formación de los planos, January 29, 1895, AMO, año 1895, caja 214, Fomento. Reiterated six years later: Circular 5, Sección de Gobierno, April 30, 1901, *CLEV—1901*, 9–10. On CGE mapping of cantons, see Memorandum: Sobre la Carta particular del Estado de Veracruz levantada por la Comisión Geográfico-Exploradora, n.d., ACGE, exp. 4, exp. de Veracruz. For previous attempts to gather information for the CGE, see J. R. Espinosa to los Presidentes Municipales y demás autoridades del Estado a quien fuere presentada, October 20, 1890, CFD, fondo 621, carpeta 1, leg. 9, doc. 2.

57 Circular num. 28—La expedida por la Jefatura Política solicitando los datos que necesito la Comisión Geográfico-Exploradora de los límites de cada municipio y copia de los títulos en que se base la posesión de cada uno de ellos, enclosure in Jefe político del cantón de Orizaba, May 4, 1895, AMO, año 1895, caja 210, Sección de Ejidos.

insignificant data."[58] This time, with their own money at stake, Vera-cruzano officials appealed to the "enlightened and patriotic fervor" of their citizenry to encourage compliance and made a special appeal to the owners and managers of rural estates to cooperate fully with the survey-ors. Problems persisted however. Apathy or resistance among hacen-dados and rancheros forced the state government in 1901 to reissue the order that they "show [the surveyors] their maps of their respective properties."[59] Meanwhile, the mayor of Orizaba was unable to gather the requested material. Local chronicler Joaquín Arroniz, in the pro-cess of writing a history of the city, had borrowed many of the docu-ments in the municipal archive and refused to return them until he was finished![60]

Still, many pueblos complied. Exactly how the prefects managed to ensure compliance is unclear, although there is little evidence of out-right coercion. It may be the case that pueblo authorities by this time saw important potential benefits in having their boundaries codified in a state map, particularly given the state's increasingly overt attempts to divide communal lands and the concomitant proliferation of boundary disputes. Regardless, by June 1895 the CGE had been inundated with material from villages and towns from around Veracruz. The corpus of material is remarkable. Pueblos submitted beautiful copies of their pri-mordial titles, some more than three hundred pages long, filled with de-tailed discussions of the founding of the pueblo, its historical rights, land conflicts, boundary rituals, and the like.[61] Most of the documents narrated, in minute detail, how a community had acquired the land it now purported to possess. Some pueblos submitted maps along with, or

58 Díaz, "Informe sobre el estado actual de la cartografía," 475–78.

59 Circular 5, Sección de Gobierno, CLEV—*1901*, 9–10. Cambrezy and Mar-chal, in their study of the changing geography of land tenure in central Veracruz, relied upon maps commissioned by hacendados in their reconstructions. Hacen-dados chartered such maps to facilitate buying and selling lands, to ensure legacies and the division of property, and to ensure fair property taxation. Still, the reitera-tion of requests directed exclusively at large landowners, and the relative sparsity of hacienda and rancho maps in the CGE archives, suggest they were not prone to share such materials with the CGE. See Cambrezy and Marchal, *Crónicas de un terri-torio fraccionado*, chap. 2.

60 Luis Echegaray, Alcalde Municipal de Orizaba, to Jefe Político del Cantón de Orizaba, June 12, 1895, AMO, año 1895, caja 210, Sección de Ejidos.

61 See ACGE, exps. 3, 5, 7, 8, and 13.

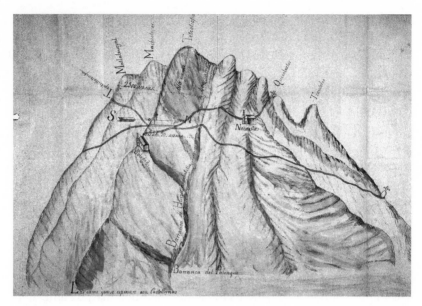

FIGURE 14. All the documents in their power: Anonymous copy of undated map of La Soledad, Veracruz, 1895. Courtesy of the Mapoteca Manuel Orozco y Berra, Mexico City. Photograph by Carmen H. Piña.

in lieu of, their documents. These maps tended to be copies of early co-lonial maps, such as the one sent in by the mayor of Soledad Atzompa (figure 14). Maps such as these were valuable documents attesting to the village's historical longevity and legal validity. They plotted out a his-torical, as much as spatial, location. Titles and maps were more than af-firmations of a right to certain lands, waters, and forests; they were a community's patrimony. Little wonder that when the authorities of Acultzingo submitted a copy of the municipality's primordial titles and 1559 map, they wrote in the frontispiece that such materials constituted "all the documents in our *poder*."[62]

"Poder" means both possession and power: such definitional confla-tion is revealing. For pueblo authorities such documents, cared for and guarded in local archives, were not only items to be possessed but power to be wielded. They understood, after generations of interaction with colonial courts, that a single title, a crumbling map, held the key to ter-ritorial and political integrity. Such a reading may sound like a leap of the imagination and such a documentary interchange too mundane to

62 ACGE, exp. 8.

merit close attention. Yet it was precisely such material that Emiliano Zapata (Mexico's most famous peasant revolutionary) ordered buried in a strongbox under the floor of the village church before committing the village to revolution. "I'm bound to die some day," he purportedly stated, "but my pueblo's papers stand to get guaranteed."[63]

While of nearly sacred value to villagers, such material was of limited practical value to surveyors attempting to delineate municipal and village boundaries. Colonial maps lacked keys for interpretation and were populated by markers and images of local importance. Titles referred to community boundaries as general areas some six hundred varas from the door of the central church and in exclusively local terms. These were not items that could be easily assimilated into a broader pattern of political geography. The difficulty of interpreting and using such material, of fitting it into a state-devised grid, compelled the CGE to send out two years later a literal grid—a "cuadro"—for municipal authorities to complete.[64] In the meantime, CGE officials were selective about what images they chose to trust. Along with their 1559 map, authorities in Acultzingo also submitted a map of Holzinger's 1872 survey, most likely in the hopes that it might aid them in their border dispute with a neighboring community in Puebla. The CGE staff carefully examined, catalogued, and archived Holzinger's map, but they apparently put the rest of Acultzingo's titles and the colonial map on a back shelf along with all the other titles. There they remained, forgotten until after the revolution, when ex-CGE staff now working for the new regime sought them out again, this time explicitly *for* their historical qualities.[65]

Cantonal and municipal borders were not the only lines of interest to state administrations. State borders had never been definitively mapped and determined in the nineteenth century. Years of political reorganizations and territorial reconfigurations merely exacerbated existing boundary problems.[66] Not surprisingly, state governments worried over how CGE operations would impact the representation of their borders. If

63 Quotation from Womack, *Zapata and the Mexican Revolution*, 371.

64 Cuadros formados por la Comisión Geográfico-Exploradora remitidos por la jefatura política para que se llenen y devuelvan antes del 15 de septiembre, con objeto de perfeccionar el mapa del Estado de Veracruz, August 26, 1897, AMO, año 1897, caja 225, Sección de Estadística, exp. 13.

65 See chapter 7.

66 See O'Gorman, *Historia de las divisiones territoriales de México*.

mapping projects were ambivalent propositions for municipalities and large landowners, they could also raise concerns for state governments. Although the CGE's work promised a potential codification and legitimation of a state's borders, officials feared losing control over determining where those borders would appear on a final map. They did not want to leave the issue of political self-definition in the hands of an agency ignorant of the social and political stakes of boundary marking. As a result, surveyors could mark the limits on the final map but based solely upon geographical and historical information provided by the state government. In effect, state administrations wanted CGE surveyors to legitimate and codify *their* version of their boundaries. "The surveyors will not trace a single border line in a definitive way on the maps that are to be published, without the express and written consent of the state government," read one clause in a typical contract.[67]

The stipulations regarding borders effectively trumped the CGE's potential to resolve the persistent questions regarding state limits. They found themselves mired by state governments' conflicts with neighboring states and by conflicts between villages in neighboring states. The case of Ixhuacan, mentioned in chapter 2, is a case in point. The division of communal lands had been under way since the 1880s, yet by 1905 the community's boundaries had still not been completely determined and fixed. The entire process was complicated by the fact that Ixhuacan bordered the state of Puebla, and the state limits had themselves proven problematic. The governor's engineers failed to resolve the issue, and Dehesa eventually ordered Ignacio Muñoz, a CGE surveyor and his compadre, back to the region to determine the border line.[68]

If CGE personnel could not mark definitive boundaries premised upon field surveys and archival investigations, they were hardly ready to confirm any boundaries delineated on the basis of limited information provided by interested parties. Alvarado, upon completion of the state map

67 Contract between the CGE and the state of Chihuahua, in Enrique Creel to Coronel Primer Ingeniero Juan B. Laurencio, September 28, 1904, ACGE, exp. 4, exp. de Chihuahua.

68 Gobernador del Estado de Puebla to Gobernador del Estado de Veracruz, January 20, 1905, AGEV, RG, Tierras, Límites, exp. Ayahualulco e Ixhuacan, 1905–1914. Muñoz's involvement dated back to 1900, when Ulloa was sent to complete the land divisions in Ayahualulco. See "En la Ciudad de Coatepec a los dos días del mes de Mayo de mil novecientos," AGEV, RG, Tierras, Límites, 1902.

of San Luis Potosí in 1894, included a small warning on the image: "[T]he State limits, although marked on the map according to the information provided by the government and political authorities of the State, should not be considered definitive."[69] He was less diplomatic in his letter to the minister of development: "In the information [regarding the state border] provided by the authorities there are a multitude of contradictions."[70] Alvarado suggested that only proper fieldwork would garner satisfactory results in terms of locating the border. The case of Veracruz was more severe. Alvarado reported in May 1901 that "the map of Veracruz . . . would have already been completed if not for the annoying and protracted question of limits."[71] Undertaken in 1895, it took a full decade for the map to see the light of day.

III. DISPLAYS

While CGE surveyors were discovering "extra places," foreign audiences and investors were rediscovering Mexico. The two decades between 1890 and 1910 were golden years for Mexico's international image. Foreign and domestic writers lauded don Porfirio's achievements at bringing order out of chaos and ushering Mexico into the modern world.[72] Promotional texts by the likes of García Cubas and Matías Romero circulated alternatives to prevailing images of Mexico as dangerous and backward.[73] And Mexico's scientific and cultural achievements were, like those of all nations, carefully choreographed on the stages of the world's fairs.[74]

69 See *Carta General del Estado de San Luis Potosí.*

70 Alvarado, *Anexo de Memoria,* 223.

71 Ibid., 302. See also Memorandum: Sobre la Carta particular del Estado de Veracruz levantada por la Comisión Geográfico-Exploradora, n.d. [ca. 1906], ACGE, exp. 4, exp. de Veracruz.

72 Examples are numerous. For a sampling of how the U.S. press presented Díaz in a significantly positive light, particularly in comparison to other Latin American leaders of the time, see Johnson, *Latin America in Caricature.* For domestic praise, see, among many others, Duclos Salinas, *The Riches of Mexico;* Caballero, *Primer almanaque histórico, artístico y monumental;* and Godoy, *Porfirio Díaz.*

73 See, for example, Romero, *México and the United States;* García Cubas, *México: Its Trade, Industries, and Resources;* and García Cubas, *Cuadro geográfico, estadístico, descriptivo, e histórico.*

74 For a superb analysis of Mexico's self-representation at these spectacles, see Tenorio-Trillo, *Mexico at the World's Fairs.*

National maps were an important aspect in presenting and represent-ing the nation, and the CGE played a prominent role in the fairs. As of 1893, although the CGE had mapped and explored some 76,000 kilome-ters of Mexico's national territory, only a few sheets of the *Carta de la República Mexicana* had actually been published, in part because the gap between fieldwork and final map could run up to three years.[75] Under Alvarado, the volume of map publication increased significantly and CGE maps became standard fare at the world expositions in Chicago (1893), Atlanta (1895), and Paris (1900).[76] These maps quickly garnered awards, praise, and most important the trust of both domestic and foreign audi-ences.[77] Yet theirs were not the only maps of importance being produced in Mexico and circulated internationally. García Cubas, since publish-ing his carta general to wide acclaim in 1858, had continued to produce more refined national maps. In 1885 he published his *Atlas pintoresco e histórico de los Estados Unidos de Mexico,* a voluminous tome composed of thirteen full-color thematic maps with images.[78] He also worked on a multitude of cartographic projects for the Paris exposition in 1890.[79] The importance of such images on the world stage is captured in the po-litical sensitivities that had to be respected. When García Cubas wrote don Porfirio a missive regarding the various images he planned on sub-mitting for the Paris exposition, don Porfirio responded by telling him to not include a particular map that, "having been produced during the Imperio," might give offense.[80]

75 Díaz, *Exposición internacional colombina de Chicago,* 9; Agustín Díaz to Por-firio Díaz, May 18, 1886, CPD, leg. 11, caja 11, doc. 5036; Alvarado, *Anexo de Mem-oria,* 219; Fernando Ferrari Pérez, "Informe de los trabajos de la Comisión Geográfico-Exploradora," 39.

76 See García Martínez, "La Comisión Geográfico-Exploradora," appendix.

77 For examples, see María Rodríguez, *Perfiles del suelo Vera Cruzano,* 19; the *Mexican Mining Journal* 10 (1), 1; and the overview in Tenorio-Trillo, *Mexico at the World's Fairs,* 131–33. Municipalities in Veracruz requested copies of CGE maps by name for their schools. See, for example, Tlacotalpam: El H. Ayuntamiento de aquella ciudad solicita cuatro cartas geográficas del Estado para sus escuelas muni-cipales, September 11, 1911, AGEV, Ramo de Fomento, Sección de Geografia y Es-tadística, caja 202.

78 García Cubas, *Atlas pintoresco e histórico.*

79 See his list in García Cubas to Porfirio Díaz, February 23, 1889, CPD, leg. 14, caja 3, doc. 1315.

80 Garcia Cubas to Porfirio Díaz, February 23, 1889, CPD, leg. 14, caja 3, doc.

Meanwhile, an array of national "network maps" appeared with increasing regularity in the last decade of the century. These images portrayed Mexico as a dense assemblage of roads, telegraph lines, railroads, and steamer routes (figure 15). Propagating an image of a multiply linked nation and economy, they demonstrated to investors that Mexico had, in Engels's words, "the means of communication adequate to modern means of production."[81] Often printed with Spanish, French, and English text, and scaled in both kilometers and miles, they were seductive images for international investors and capital. A multitude of other national maps served as visual complements to the numerous works commissioned by the Mexican government and published for foreign business audiences, which extolled Mexico's prosperity, industry, and commercial potential. They were visual ciphers for the volumes of statistics produced every quarter by local, regional, and national administrations. Porfirian Mexico experienced a veritable explosion of statistical output, and maps often captured more dramatically the data compiled. In visual shorthand one could see Mexico's progress through, as García Cubas put it to don Porfirio, "the eloquence of its numbers."[82] The images thus told their own history, one that assured audiences of the Mexican nation's progression from stormy adolescence to modern maturity. Capitalist, productive, and stable, with a rich history and an even richer geography, Mexico appeared ripe for study, colonization, and exploitation. On the surfaces of network maps and in the pages of pictorial atlases, Mexico came together as a world of exotic past and capitalist future.

Although of great value, such images had their limitations. García Cubas's pictorial-descriptive maps, bound as lavish atlases, were intended for the mantels and shelves of elite households, both foreign and domestic. They were designed for bourgeois consumption rather than bureaucratic consultation, and while they inspired the imagination, they did not move the investor. The quickly multiplying array of network maps, meanwhile, were too thematic to be of broad general use and

---

1315; Porfirio Díaz to García Cubas, February 20, 1889, CPD, leg. 14, caja 3, doc. 1316; and García Cubas to Porfirio Díaz, March 5, 1889, CPD, leg. 14, caja 5, doc. 2288.

81 Quotation from Hobsbawm, *The Age of Capital*, 33.

82 The quoted text is from García Cubas to Porfirio Díaz, February 23, 1889, CPD, leg. 14, caja 3, doc. 1315.

FIGURE 15. The network nation: Francisco Calderón, *Carta postal y de vías de comunicación de los Estados Unidos Mexicanos*, 1910. Courtesy of the Mapoteca Manuel Orozco y Berra, Mexico City. Photograph by Carmen H. Piña.

lacked topographic precision. Neither set of images constituted the kinds of "precise and general map[s] of the territory" that "all cultured nations have," observed Eugenio Chavero, a senior official in the Ministry of Development attempting to drum up financial support for the CGE.[83] "It would be inexplicable," he continued, "that Mexico, having arrived at such an advanced state of civilization, lacked something so indispensable" as a precise map.[84]

Indeed, a *precise* national map had become, by the late nineteenth century, both the image and measure of a nation-state's modernity. And a precise general map was a product of scientific practice: a map grounded in the objective authority of sophisticated instruments, careful calculations, and a combination of field- and office work. Science itself was a cultural litmus test, a language of legitimacy, a means by which to distinguish modern, civilized nation-states from their supposedly retrograde inferiors. Vicente Riva Palacio and other Porfirian "wizards of progress" sought persistently, at international exhibitions, to elevate Mexico's cultural status by reference to scientific activity.[85] In Spain in 1891 for the four hundredth anniversary celebrations of the "discovery of America," Riva Palacio suggested that this was an "occasion to prove that our scientific and literary level is, if not higher, at least as high as here."[86] One of Riva Palacio's successors as minister of development, Manuel Fernández Leal, complained some years later of the "primitive maps" of the Republic produced by individuals "not competent in the scientific method."[87]

While Fernández Leal might have balked at calling García Cubas's ornate maps "primitive," they certainly did not reach Riva Palacio's scientific heights. Scientific images, like scientific authority, succeeded by conveying a sense of constraint, composure, and self-regulation. Painted images on borders, elaborate cartouches, and similar paraphernalia simply drew attention to the subjective and interested reality of the

83 Eugenio Chavero, introduction to *Memoria presentada al Congreso de la Unión . . . enero de 1883 a junio de 1885*, 1:2.

84 Ibid.

85 "Wizards of progress" comes from Tenorio-Trillo, *Mexico at the World's Fairs*.

86 Vicente Riva Palacio to Porfirio Díaz, January 22, 1891, CPD, leg. 16, caja 3, doc. 1400.

87 *Memoria presentada al Congreso de la Unión . . . 1897 a 1900*, 115–16.

map itself. The form called into question the content. The professed objectivity of a scientific map derived precisely from its presumed refusal to revel in the subjective world of art. In other words, objectivity had an aesthetic standard.

As much as they eschewed any hint of subjectivity, scientific maps also achieved their objective (and objectivity) through rhetorical means of persuasion. In part, the mapped image itself contained such persuasive devices: graticules and coordinates, for example, created an impression of objectivity. While necessary, those were not sufficient in and of themselves. How did one know that Mexico City had been correctly positioned in the grid? Or that the grid itself had been accurately determined? How could potential investors—whose successes were predicated upon predictable, homogeneous, and transparent space—come to trust the image that confronted them? How did a scientific map acquire the very adjective that conveyed and instilled confidence, authority, and trust?

The perceived veracity of the works of the CGE hung on the journals and notebooks that detailed the manner in which they were constructed. The acceptance of the image as scientific fact was directly connected to the layers of texts that fashioned the producer as authoritative, objective, and competent.[88] Even at the height of positivist confidence, scientists were aware that they needed to *write*, not merely *write up*, the results of their research.[89] Thus, Díaz and Alvarado filled pages of accompanying text for the world's fairs with elaborate descriptions of the procedures by which measurements were made, recorded, and verified: how they calculated geographic coordinates; how they reduced the potential for error in observations and calculations; how they determined longitude in multiple ways to ensure accuracy; how they measured altitude; how the maximum margin of error in any calculation fell well within international standards; and how all measurements were made by trained personnel with recognized brand-name instruments.[90] These

88 See Lenoir, "Inscription Practices and Materialities of Communication," and Schaffer, "The Leviathan of Parsonstown." See also Shapin and Schaffer, *Leviathan and the Air Pump*; Burnett, *Masters of All They Surveyed*, chap. 3.

89 I am indebted here to the fascinating discussion in Mermin, "Writing Physics," for this distinction.

90 See, for example, Díaz, "Catálogo de los objetos." The Mapoteca Orozco y Berra holds many of these journals. Angel Anguiano, head of the Mexican Geode-

tomes of lexical persuasion, filled with tedious calculations and records of measurement, were the talismans of the map's scientific authority, buttressing its facticity by drawing attention to the process of its inscription.[91] Esoteric equations and symbols, the hard logic of calculus and geometry, furthered the artifice of objectivity and transparency by purporting to render visible the process of the map's formation. Mathematics, after all, allowed no opinion. If the fieldwork of the surveyors served as a basis for empirical reliability, the books of Díaz and Alvarado attested to the other half of objectivity: procedural precision.[92] Such textual matter demonstrated Mexico's comfort with the advanced practices of cartographic science and fended off any suggestions of scientific malfeasance.

Such codes were particularly important given Mexico's structural circumstances relative to France, Britain, and the United States. They constituted a kind of aggressive defensiveness, as both Díaz and Alvarado were quick to point out the CGE's financial and technical shortcomings.[93] They had not carried out primary order geodesic surveys of the kind undertaken in France and British India, and thus the texts of the CGE's catalogues that traveled the exhibitionary circuit inevitably offered a number of caveats for the viewers.[94] Alvarado warned them not to misconstrue Mexican pride for arrogance: "We would never presume that our work would compete with the grand geodesic operations undertaken in the Old World."[95] Díaz had been even blunter. Mexico, he wrote, occupied a "level of cultural sophistication inferior to that of France and other nations participating in the Exposition."[96] Still, Díaz's humility toward the old world did not preclude a small amount of hemispheric hubris as he sought to convince "the countries of

---

sic Commission, in a series of essays on Mexican cartography published over the course of two years, critiqued García Cubas's images for the absence of this kind of data. Anguiano, "Cartografía Mexicana," 168–69.

91 Cf. Latour, *Science in Action*.

92 On empirical reliability, see Daston and Galison, "The Image of Objectivity," esp. 82.

93 I am indebted to Peter Dear for suggesting seeing this as a kind of "aggressive defensiveness."

94 On geodesic surveys and CGE methodology, see chapter 4.

95 Alvarado, *Comisión Geográfico-Exploradora . . . París en 1900*, 3–4; Alvarado, "Catálogo de los objetos . . . Exposición de Londres," ACGE, paq. 162.

96 Díaz, "Catálogo de los objetos."

South America" to adopt the Mexican "manner of proceeding," which offered a comfortably functional combination of speed and relative precision.[97]

If the textual scaffolding proved of paramount importance, so did the facade of the image. Scale, symbols—an entire map iconography—were tailored to the audience, designed to endow the image with a veneer of rationality and authority. Just as a host of criteria had been developed regarding the configuration of knowledge during the creation of field itinerarios, so too did an infinite number of potential ways to construct the final maps have to be narrowed down to a single, uniform cartographic iconography. Díaz took charge in detailing how the finished map would appear. His lengthy deliberations reveal that such concerns were hardly trivial. They constituted important dimensions to how the nation-state would be portrayed and viewed.

Díaz was especially concerned with the international audience of governments and investors who might use the maps. International compatibility of the image thus became an important consideration. For example, Díaz chose the recently adopted international scale of 1:100,000 for the master map of Mexico.[98] Similarly, when he divided the country up into a series of square segments, each assigned a letter from the alphabet, he excluded the *ñ* and the *ll* to make the images more accessible to an English-language audience.[99] In developing the symbols that would grace the CGE's maps—denoting everything from archaeological ruins to functioning mills and congregaciones to state capitals—he consulted the French code of standardized map symbols.[100] Yet Díaz also expressed concern that the CGE not be construed as simply mimicking European cartography, and he altered the symbols to avoid anything that smacked of imitation.[101] Díaz obsessed over symbolism in other ways. He insisted, for example, that individual map sheets be proportioned in such a way as to consistently give Mexico City, as well as the *M* in *Mexico*, central prominence.[102] Díaz understood the powerful, if subtle, effect such

97 Ibid.

98 Noriega, "Los progresos de la geografía de Mexico," 264–69.

99 Díaz, "Catálogo de los objetos."

100 Díaz, untitled notes on signs and symbols, ACGE, carpeta 106.

101 Ibid., and Alvarado, *Comisión Geográfico-Exploradora. . . . París en 1900,* 25–26.

102 Díaz, "Catálogo de los objetos."

iconographic care could have in articulating an image of a stable, centralized nation-state.

In sum, the politics of the image fostered an image of politics. Iconographic subtlety, combined with the certitude of scientific methodology, conveyed an impression not only of the land but of the regime that mapped and managed it. Maps were not just mediums for the transmission of evidence; they were evidentiary in and of themselves, "visual metaphors" for Porfirian Mexico.[103] Their ordered surface suggested a corresponding political, economic, and moral order, and integration in the state itself. Their exterior implied concordant degrees of stability and predictability in the worlds of politics, commerce, real estate, and society, even if such stability and integration proved more myth than reality on the ground.[104] What better visual complement to scientific politics (with its neutral slogan: *mucha administración y poca política* [a lot of administration and not too much politics]) than the scientific map? What better instrument for, and image of, administered, rational rule? In effect, scientific practices melded with political theory in a teleology of order and progress, the endpoint of which was the modern, capitalist state.[105] As this teleology found both expression and actualization in the maps of the CGE, Mexican cartography became a quintessentially *political* science.

CONCLUSION

Image isn't everything. Appearances are deceiving. A host of tensions and conflicts sat uncomfortably close to the orderly surface and, in 1910,

103 "Visual metaphor" comes from Edgerton Jr., *The Renaissance Rediscovery of Linear Perspective*. On the politics of "the image of the nation," particularly during the Porfiriato, see Tenorio-Trillo, *Mexico at the World's Fairs*. Mexican cities were prime sites for creating an image and aesthetic associated with modernity. For the case of Mexico City, see Tenorio-Trillo, "1910 Mexico City"; on provincial capitals such as Mérida and Oaxaca City, see Wells and Joseph, *Summer of Discontent, Seasons of Upheaval*, chap. 4; and Overmyer-Velázquez, "Visions of the Emerald City."

104 A recent collection makes clear that Porfirio's power was never as "omnipotent" and all-encompassing as has often been believed. See Falcón and Buve, eds., *Don Porfirio Presidente . . . , nunca omnipotente*.

105 The masking of power and political interests with the facade of science—including scientific renderings of abstract space—is a defining feature of the mod-

cracks appeared in the facade. Their rapid proliferation brought an end to Porfirian order and progress. It also brought an end to the CGE, although its passing was more protracted.

The CGE was a Porfirian institution, and when don Porfirio's rule collapsed, its demise was imminent. Francisco Madero actually maintained the agency, enlisting its director, Angel García Peña, as his new minister of war. The CGE persisted also under Victoriano Huerta's short-lived regime, not a surprising fact given Huerta's long and intimate relationship with the institution. Indeed, Huerta counted on the support of former CGE colleagues, such as Ignacio Muñoz, in his rise to power.[106] But with Huerta's fall and the abolition of the federal army, the CGE finally disappeared from the Mexican landscape. So too did many of the CGE's more senior personnel.

The very thing that made the CGE such an effective tool of rule for don Porfirio also ensured that it could never outlast the man himself. Committed Porfiristas, mostly from the federal army, populated the upper echelons of the agency. They came from well-to-do backgrounds and often owed their financial clout and political power to the good graces of don Porfirio himself. A cursory sampling of the networks among CGE personnel in Veracruz rings like a Who's Who of Porfirian high society: Ignacio Muñoz and Félix Díaz, both nephews of don Porfirio, worked as CGE personnel in Veracruz for most of don Porfirio's tenure. Muñoz was the compadre of Veracruz governor Teodoro Dehesa, and Dehesa's son had married the daughter of CGE surveyor and director Angel García Peña.[107] Félix Díaz wed Isabella Alcolea, daughter of Leandro Alcolea, Sr., a Veracruzano of high standing and member of the Porfirian Chamber of Deputies.[108] No wonder that by late 1914, with Huerta out of the picture and Villa and Zapata stepping into it, prominent members of the CGE joined Governor Dehesa and an array of Porfirian associates packing the port of Veracruz, primed for departure to

---

ern bourgeois state. See Poulantzas, *Political Power and Social Classes*, and Lefebvre, *The Social Production of Space*. On modernity and the emergent connection between scientific ideas about nature and political ideas about society, see Toulmin, *Cosmopolis*, esp. chap. 3; Foucault, *The Archaeology of Knowledge;* and Foucault, *The Order of Things.*

106 Pasquel, *La revolución en el Estado de Veracruz*, 2:11.

107 Ibid., 1:38–41.

108 See Henderson, *Félix Díaz, the Porfirians and the Mexican Revolution*, 4.

New Orleans, Havana, or beyond.[109] Dehesa and his family eventually ended up in Jacksonville, Florida, where they anxiously awaited news regarding their possible return.[110] Other cge officials followed soon after. Officers in Venustiano Carranza's Constitutionalist Army captured García Peña and his son-in-law in early 1915 as they attempted to board a Spanish liner bound for an unknown destination.[111] Francisco Canóvas, who had left the cge in 1902 to become prefect of the canton of Misantla, also fled into exile in 1915. Canóvas served as prefect until May 20, 1911, when he was deposed by a force of some 125 revolutionaries entering Misantla.[112] He soon picked up surveying work dividing communal lands in the southern district of Cosamaloapan.[113] But in 1915, confronted by the impending U.S. withdrawal from Veracruz, the real threat of draconian "revolutionary justice," and accusations of counterrevolutionary activity, he fled to Havana, Cuba.[114]

An affiliation with the cge alone was not enough to condemn one to the potential harshness of revolutionary justice, even if some carrancistas saw all ex-*federales* as "irreconcilable enemies of the Consitutionalist cause."[115] Some were: a carrancista counterespionage report noted that

109 Knight, *The Mexican Revolution*, 2:179.

110 Angel García Peña to Venustiano Carranza, June 15, 1915, cvc, fondo 21, carpeta 42, leg. 4588, doc. 1.

111 Angel García Peña to Venustiano Carranza, February 16, 1915, cvc, fondo 21, carpeta 28, leg. 2894.

112 See Correspondencia con la Jefatura Política del Cantón, durante el año de 1911, amm, caja 1911; and Copia de los mensajes depositadas en esta oficina en la Telegráfica local durante el año actual (May 9–19), amm, caja 1911, exp. Sección de Hacienda, no. 50. Canóvas mentions his length of tenure in "Refutación hecha por el Ingeniero Francisco Canóvas," fcp.

113 Alcalde Municipal de Cosamaloapam to Francisco Canóvas, March 14, 1914; and Bases del contrato que deberá celebrar el H. Ayuntamiento de Cosamaloapam con el Ciudadano Ingeniero Francisco Canóvas para la medición y planificación del ejido y fundo legal, May 19, 1914, fcp.

114 Accusations against Canóvas, as well as his brother Sebastián, appear in Luis G. Morales to General Cándido Aguilar, August 14, 1915, cvc, fondo 21, carpeta 48, leg. 5346. His place of exile is revealed in his collected letters in fcp. On the carrancista campaigns against former porfiristas and huertistas, see Knight, *The Mexican Revolution*, 2:180–84, 443–44.

115 Dario Hernández to Venustiano Carranza, May 28, 1916, cvc, fondo 21, carpeta 80, leg. 8854. See also Eduardo de Gortari to Cándido Aguilar, March 23, 1916, cvc, fondo 21, carpeta 71, leg. 7758.

Ignacio Muñoz continued his counterrevolutionary activities in the port of Veracruz.[116] Yet other CGE officials not only escaped persecution but found employment in the new administrations. CGE artists Carlos Rivera (a highly-regarded artist from Xalapa) and Carlos Olivares found work in the Local Agrarian Commission of Veracruz and the Department of Public Works.[117] And as the carrancistas solidified control in the waning years of the armed phase of the revolution, former enemies were allowed to return. García Peña, having attained his liberty, eventually managed to broker permission for Teodoro Dehesa and his family, now in Havana, to return. Although García Peña had stressed that Dehesa had "always been an enemy of the *científicos*," the regime's conciliatory attitude may have rested just as much on their knowledge that a gravely ill Señora Dehesa wished to die in Veracruz.[118] Also returning from exile in Havana was Francisco Canóvas, who soon found work as a surveyor in the Comisión Nacional Agraria, creating ejidos.

116 Coronel Nicolás Velarde, "Informe de servicio de contra-espionaje, ejército constitucionalista," October 27, 1916, CVC, fondo 21, carpeta 100, leg. 11426.

117 See the correspondence in AGEV, RF, Sección de Geografía y Estadística, caja 202, año 1923, Asunto: Suplicándole que devuelve a este Gobierno el ARCHIVO que fue facilitado en calidad de préstamo a la Comisión Geográfico-Exploradora que estuvo instalada en la Capital del Estado, January 11, 1923.

118 Angel García Peña to Venustiano Carranza, June 15, 1915, CVC, fondo 21, carpeta 42, leg. 4588, doc. 1; Raúl Dehesa to Angel García Peña, October 23, 1917, CVC, fondo 21, carpeta 118, leg. 13384. The "científicos" were adherents of positivism who occupied positions in the upper echelons of Porfirio Díaz's administration. They were often associated with the more elitist, and nefarious, aspects of don Porfirio's rule: hence García Peña's efforts to vindicate an old Porfirista like Teodoro Dehesa by claiming him to be "an enemy of the científicos."

CHAPTER 6

Fluvial Confusions

In 1905 the mayor of Las Minas, a village located deep in the barranca (canyon) of Zomelahuacán in the western highlands of Veracruz, received distressing news. Albert Schuler, a mine owner and former resident, was attempting to wrangle a concession to the water of a river that coursed along the town's banks. With a hand-drawn copy of a map from the newly published plates of the Comisión Geográfico-Exploradora, carefully modified with sharp lines and ambiguously inscribed names, he had captured the ear of the federal Ministry of Development. The mayor immediately called an emergency session of the village council. The following day, he had a letter and four documents, meticulously copied from the local archives, sent to the prefect.[1] There ensued years of documentary warfare between the council members and Schuler, a war waged with maps and names.

The crux of the conflict revolved around the names of the waterways striating the deep valley. As the Porfirian administration attempted to bring waterways under federal dominion, confusions arose as to which rivers were under federal jurisdiction and which remained under local control. More accurately, confusions arose as to which river had which name. Federal agencies may have emphatically determined, for example,

---

1 This reconstruction and the bulk of the following chapter is based upon various expedientes in the Archivo Histórico del Agua [AHA], Fondo Aprovechamientos Superficiales [AS]. The main sources are caja 1168, exp. 16333; caja 4267, exp. 57049; caja 4268, exp. 57063; caja 4516, exp. 59824; caja 4521, exp. 59912; caja 4528, exp. 60048; caja 4532, exp. 60085; caja 4532, exp. 60104; and caja 7544, exp. 60301. For the sake of economy, hereafter all references to documents from these will be by date and expediente only.

that the Minas River fell under federal jurisdiction, but in an area striped with waterways, which river *was* the Minas River? At points of confluence, which waterway denoted the continuance of the Minas and which its tributary? Both Schuler and the municipal council sought to ensure that *their* names of rivers and tributaries prevailed. To do so, both appealed to the Ministry of Development, which, in conjunction with the Ministry of Communications and Public Works, assumed the task of determining which images and place-names would be considered correct.

"If names be not correct, language is not in accordance with the truth of things; if language is not in accordance with the truth of things, affairs cannot be carried on to success."[2] Confucius's dictum, although concerned more with conceptual than topographic clarity, still succinctly summarized the situation for both Schuler and the municipal council. The names that prevailed, those deemed correct, would determine whose affairs would sink and whose would swim. In this instance, the pen truly prevailed over the sword. It was parchment—registrations, deeds, titles, and maps—that each party raised in their clutched fist. And as the conflict unfolded, blurred lines were sharpened, new images were drawn, and place-names became tools of sabotage.

I. POWER LINES

In 1905 Albert Schuler celebrated his sixtieth birthday. A U.S. citizen, Schuler had come to Mexico some years previously, settling in the canyon of Las Minas where he worked as a self-described *minero* (miner). By this he did not mean that he labored in the still darkness of the earth but rather that he owned and ran a number of mines. In fact, he had accrued, over the course of the previous half-decade, a substantial amount of property in the canyon, mostly in the immediate vicinity of the pueblo of Las Minas.[3] Most recently, he had purchased the extensive holdings of

2 Soothill, *The Analects of Confucius*, 608.

3 Schuler's age is provided in Testimonio de la escritura de cesión onerosa formalizada por el Señor Don José de Prida como heredero y albacea del Señor su padre Don Francisco M. de Prida a favor de los Señores Don E. Guillermo Vogel y Don Alberto Schuler, AHA, exp. 59824. On his citizenship and residency, see Schuler to Ministro de Fomento, December 16, 1905, AHA, exp. 57063, f. 29. Schuler bought most of his land and mines between 1901 and 1905; see the collected documents in AHA, exp. 59824.

the Compañía Minera de Zomelahuacán. Although he had lived in Las Minas for quite some time, by 1905 he had left the valley for the comforts of the Grand Hotel in Mexico City, where he could more easily meet and plan a comfortable future with his business partner, E. Guillermo Vogel, a German investor and businessman with substantial business experience in Mexico.[4] Together, they purchased lands and dilapidated *haciendas de beneficio* (processing plants) in the canyon and formed the Mexicana Mining and Smelting Company, which Schuler claimed to have over U.S.$2,000,000 in capital, to mine the veins of the canyon.[5]

The mines of Las Minas were a mixed proposition. Copper, lead, coal, and gold existed in indeterminate but tantalizing quantities, and some authorities reckoned the iron deposits in the region were a veritable mother lode awaiting extraction by an enterprising individual.[6] Such enticing allegations were countered by the blunt statements of travelers such as Alexander von Humboldt, who, having found the mines of Zomelahuacán "virtually abandoned," dismissed the Intendency of Veracruz as a place with no mining potential.[7] Two decades later the English, during their investment boom in Mexico, established the Mexican Mining Company there and extracted enough gold to produce commemorative watches inscribed with the words: "Mexican Gold, from the mines of Zomelahuacán. London, 1827."[8] Yet, the very year before the watches were minted, the company pulled out of Veracruz, seeking better times in Oaxaca. For the next half-century wealthy businessmen from Xalapa

4 Vogel worked in the 1880s as a representative for the important Compañía de Navegación del Pacífico, which ran a major shipping operation between Mexico, the United States, and China. See L. Larraza, Salvador Malo, and E. Guillermo Vogel to Porfirio Díaz, March 19, 1886, CPD, leg. 11, caja 6, docs. 2891–92.

5 Schuler's grandiose claims are dubious. U.S.$2 million in capital would have made the company one of the major players in the mining industry in all of Mexico, yet I have found no mention of Schuler nor his company in any of the major mining publications regularly circulated at the time.

6 García Cubas, *Atlas mexicano geográfico y estadístico*, carta 6, Veracruz, 2. On iron, see Sección de Catastro: Datos rendidos por las Municipalidades del Cantón de Jalacingo para la corrección de la Carta General Geográfica del Estado, June 12, 1888, ACGE, exp. 3 [hereafter: "Sección de Catastro: Datos rendidos"]; and Capilla "Los yacimientos de fierro de Tatatila."

7 Quotation from Poblett Miranda, ed., *Cien Viajeros en Veracruz*, 2:116.

8 Fenochio, *Informe acerca del Mineral de Zomelahuacán*, 18.

and Teziuhtlan owned the mines, working them only intermittently. New mining and tax codes passed in the 1880s revived foreign interest, and a new foreign enterprise, the Quinby Appleton Company, invested in the region, owning and working the largest and most important mine in the region (the San Anselmo) as well as a host of others.[9]

Regardless of the tax breaks and their monopoly of the best mines, the company did not last long. By the last years of the century it had sold off its holdings to a number of individuals, including Francisco de Prida, a Veracruzano who formed the Compañía Minera de Zomelahuacán. While the minister of development reported in 1897 that Prida's venture was each day getting "more satisfactory results," the future of mining in the canyon appeared in doubt.[10] By now potential investors realized that the mines would never live up to the billing, proffering miniscule amounts of ore compared to the mines of the central plateau and the north.[11] More problematic still was the nature of the canyon. Described by Orozco y Berra as a "narrow pit enclosing a few human beings" who lived with only some seven hours of sunlight a day, the constricted canyon did not easily yield its product.[12] The expense of transporting the ore out of the canyon and to regional centers, rather than the quantity or quality of ore extracted, constituted the most pressing problem for mine owners. The closest railroad passed through the small town of Las Vigas, some 1,550 meters above the floor of the canyon. The difficult and dangerous ascent required the use of an array of switchbacks some twelve kilometers in distance. The dangers were severe enough that the Appleton Company, in 1883, had to pay to have mail delivered to their employees in the canyon.[13] By the turn of the century, when even the inhabitants of the rugged sierra of Zongolica could count on mail by horse, those in Las Minas still had to wait for delivery on foot.[14] Orozco y Berra suggested that in fact "it would be mad-

9 Ibid., 25–47. On the mining codes and mining laws, see Bernstein, *The Mexican Mining Industry*, 18–19 and Nava Oteo, "La minería bajo el porfiriato."

10 *Memoria presentada al Congreso de la Unión . . . 1892 a 1896*, 89.

11 See *Memoria presentada al Congreso de la Unión . . . 1897 a 1900*, 550–51.

12 Quoted in Fenochio, *Informe acerca del Mineral de Zomelahuacán*, 10. See also "Sección de Catastro: Datos rendidos."

13 Fenochio, *Informe acerca del Mineral de Zomelahuacán*, 16.

14 See "Carta Postal de la República Mexicana," Hoja 21, Estado de Veracruz, July 1, 1901, MOB, CGV, varilla 2, no. 6814.

ness" to try to descend into the barranca by horse "regardless of the abilities of the rider."[15] And it appeared it would be madness for a mining company to persist in such dismal conditions: García Cubas reported that the costs of removal and transportation were so high that little or no profit remained when all was said and done.[16]

No wonder so many companies had come and failed. The remains of small processing plants and mills littered the landscape, their tiled roofs broken, beams rotted, and facades crumbling under the crush of constant rains and lush vegetation. Decrepit and forlorn, silent sentinels of mines forsaken for better prospects, their walls served only to magnify the dull roar of the rivers that coursed some meters away. Yet therein lay the solution. Cascading down the precipitous cliffs and striating the enclosed valley, these rivers held the promise of revitalization. This year-round fluvial abundance—the combined product of the snowmelt in the higher peaks, the springs around the Cofre de Perote, and abundant rainfall—held the key to mining's resurgence in the valley in the 1890s. With the rise in hydroelectric technology, the force and volume of the various channels could generate the power necessary to run every aspect of the mining process—hydroelectric turbines for lights, drills, pumps, hoists, trams, crushers, and air compressors—all of which made the mines more efficient and substantially reduced costs.[17] Indeed, Juan Fenochio, a scientist working for the Ministry of Development, observed on his excursion through the canyon in 1883 that the rivers in the region had "enough water year round to power machinery."[18] They could power a smelter, thereby permitting one to reduce the ore to metal in the valley and thus substantially lessen the weight and size of material to be hauled out. They could also help power a cable system to take the ore or metals out of the valley and up to Las Vigas.[19] The rivers were, by all accounts, virtual power lines.

Albert Schuler and Guillermo Vogel understood well enough that without rights to the waterways, ownership of the mines amounted to

15 Quoted in Fenochio, *Informe acerca del Mineral de Zomelahuacán*, 10.

16 García Cubas, *Diccionario geográfico, histórico y biográfico*, 5:513–15.

17 On hydroelectric technology and its impact on mining, see Bernstein, *The Mexican Mining Industry*, 42–45.

18 Fenochio, *Informe acerca del Mineral de Zomelahuacán*, 17.

19 As one company had begun to attempt by 1905. See Capilla, "Los yacimientos de fierro de Tatatila."

little. Beginning in 1902 they, along with Vogel's son Pablo, filed a series of petitions to the federal government for concessions to the waters of various rivers in the region.[20] Petitions such as these were required under the guidelines of a federal decree of 1894, which specified the steps an applicant had to take for a water concession. As well as an official solicitation (duly published in the *Diario Oficial* if approved), applicants had to include maps and descriptive reports of the region in question; a description of the works to be undertaken within a determined time frame; and agree to cover the costs of an engineer, appointed by the federal government, to inspect these works.[21] This decree was the second of a series passed by the Porfirian government as part of its attempts to assert national control over hydraulic resources, part of a broader process that Francois-Xavier Guerra has described as, perhaps somewhat too emphatically, the "extinction of local autonomy."[22] The first had been the 1888 Ley de Vías Generales de Comunicación, aimed at arrogating power over water from local municipal councils, who had inherited the rights granted to communities under the colonial government.[23] Limited in aim and cautious in approach, the law of 1888 delineated the conditions under which a river or lake would fall under federal jurisdiction: if it served as a dividing line between two states, had a significant enough volume of water to justify being considered a resource of national importance, or was navigable or *flotable* (suitable for floating or transporting items of commercial value).[24] The law did not give the federal

20 Albert Schuler to Ministro de Fomento, February 28, 1902, AHA, exp. 60104; Guillermo Vogel to Secretario de Fomento, February 28, 1902, AHA, exp. 57049.

21 See Leyes sobre aguas de jurisdicción federal de 6 de junio de 1894, in *Memoria presentada al Congreso de la Unión . . . 1892 a 1896*, 402–3.

22 Guerra, *México: Del antiguo régimen a la revolución*, 1:281–82.

23 The best introductions to the history of water rights and water policy in Mexico are Aboites, *El agua de la nación*, and Kroeber, *Man, Land and Water*. On ayuntamiento control over water, see Aboites, *El agua de la nación*, and the essays in Suárez Cortez, ed., *Historia de los usos del agua en México*.

24 See Leyes sobre aguas de jurisdicción federal de 5 de junio de 1888, in *Memoria presentada al Congreso de la Unión . . . 1892 a 1896*, 401–2. The Constitution of 1857 decreed that rivers which served as "vías generales de comunicación" were to be considered under federal jurisdiction, but, as one official noted in 1885, there were no general laws in existence regarding water and water rights, although the secretary of development had one "pending in the Senate." As a result, for the Dis-

government actual proprietary rights over these waterways, only juris-diction.

These were tentative steps designed to gradually centralize adminis-tration over proprietary rights of waterways without radically alien-ating entrenched and powerful local councils and *caciques*, who had en-joyed years of institutionalized rights to determine the use of their respective waterways. At the same time, the laws served to gradually po-sition the federal government as arbiter and final authority on questions over water rights. For example, although the law of 1888 did not give the government explicit right to grant concessions to individuals or com-munities, the administration began to do so anyway.[25] The decree of 1894 legalized this de facto practice by endowing the federal govern-ment with the power "to make concessions to individuals or companies for the best use of federal waters, in both irrigation and for its potential application to diverse industries."[26] While this only initially applied to the federal district of Mexico City and federal territories, by 1902 new legislation extended this control nationwide, making the federal gov-ernment the primary landlord of many waterways in the country. These legislative maneuvers reflect not only the government's recognition of the incredible potential of Mexico's waterways for both agricultural and hydroelectric production, but also that the state should be the legitimate title owner to such waterways.

Not surprisingly, however, federal officials had little reliable in-formation about the vast majority of waterways in the country, except for sizeable rivers such as the Pánuco, Papaloapan, Nazas, and the Río Bravo. Officials had no idea how much water flowed in the various riv-ers, how much of the flow might still be uncommitted, nor of the status of preexisting rights granted during or even since the colonial era. Laws

---

trito Federal (D.F.) and territories, the government relied upon the "antiguas leyes Españolas"; states controlled questions of water rights through their own specific civil codes, which, almost uniformly, were the same as that of the D.F. See the exchange between William Ham Hall, Office of the State Engineer, California, and an anonymous Mexican official, October 31, 1885, CPD, leg. 10, docs. 9608–10. More generally see Kroeber, *Man, Land and Water,* and Aboites, *El agua de la nación.*

25 See Aboites, *El agua de la nación,* 82–84; and Kroeber, *Man, Land and Water,* 170.

26 Quoted from Aboites, *El agua de la nación,* 85.

and decrees meant little if the government lacked the knowledge and information necessary to specify, delineate, and grant concessions. When local authorities argued, as did the municipal council members of Las Minas, that they had had in their possession "these waters since the time the region was made a municipality and was granted lands, since the time of the viceroys," the federal government was in no place to argue.[27]

Thus, in the years following the promulgation of the law of 1888, the Porfirian bureaucracy took the first steps to compile the necessary information. In 1889, minister of development Carlos Pacheco put out a comunique to all governors and prefects requesting information on the rivers in their respective jurisdictions. Such information included where a river originated; whether or not the river served as a limit between other states; if the river was navigable; the location and names of its tributaries and the populations located along its banks; and the river's approximate length and the names by which it was known.[28] The government also pursued production of a hydrographic map, which appeared in 1897, and charged the CGE with mapping waterways.[29]

Such intermittent efforts were only partially successful at best. A short review of some of the confusion that persisted regarding names and rivers, just in the canyon of Zomelahuacán alone, is revealing. When Pablo Vogel (Guillermo's son) applied for rights to a number of waterways in the canyon in 1902, José de Prida opposed his petition, claiming in fact that what Vogel described in his solicitation did not conform to the map Prida had in his possession. Prida argued that the names Vogel used to refer to certain rivers did not match those found on the map; as a consequence, he could not determine for sure if the petition conflicted with his concessions and interests.[30] Two years later, when an individual petitioned for rights to a river in another part of the valley, he did not name the waterways at all, claiming that no one really

27 Alcalde Municipal de Las Minas to Jefe Político de Jalacingo, transcribed in Jefe Político to Secretaría de Fomento, January 3, 1906, AHA, exp. 57063, ff. 32–34.

28 Carlos Pacheco, Secretaría de Fomento, April 10, 1889, ACGE, carpeta 113.

29 Noriega, "Los progresos de la geografía de México." See also Angel García Peña to don Luis Terrazas, Gobernador del Estado de Chihuahua, October 27, 1903, ACGE, exp. 4, exp. de Chihuahua.

30 José de Prida to Ministro de Fomento, May 28, 1902, AHA, exp. 57076.

seemed to know what the names were.[31] If at times solicitants included no names in their petitions, in other cases they included a parenthetical containing every possible name the river had been given.

Such confusion, on the one hand, perpetuated the chaos that characterized the process of determining use rights and granting concessions. At the same time, as disputants increasingly sought recourse from the federal government, the very process of clarification legitimated state power and provided federal officials with the images, documents, and information necessary to imprint their authority. In other words, authoritative maps were produced out of disputes and conflicts as much as they resulted from planned surveys. Disputants were hardly ignorant of this simple fact, and, understandably, in such circumstances images assumed remarkable power. A representational recasting of the landscape could prophesy its juridical transformation, with serious consequences for peoples' lives and livelihoods. Accordingly, adversaries waged fierce paper battles as the state attempted to fix the fluid terrain of the Mexican countryside.

II. FLUID TERRAIN

In 1905 Schuler requested a concession from the Ministry of Development "to use, as a power source and for the processing of metals, the waters of the Minas River in the quantity of 10,000 liters per second."[32] Unlike his previous petitions, Schuler included an additional item for the state's perusal—a map: "In order to make my request clear, I attach a sketch of the locale of Las Minas, taken from the work of the Geographic-Exploration Commission [and] according to the general map of the state, in which the concession I request appears on the Minas River with a blue line."[33] In the croquis, Schuler highlighted the solicited waterway, tracing its trajectory through the maze of local tributaries and rapidly changing names (figure 16).

31 Hermelindo Lechuga to Secretaría de Fomento, November 17, 1904, AHA, exp. 59914.

32 Schuler to Ministro de Fomento, October 29, 1905, AHA, exp. 57063, f. 3r–v.

33 Albert Schuler to Secretaría de Fomento, October 29, 1905, AHA, exp. 57063, f. 3r–v. The cartographic sheet of this region of Veracruz had been published in 1903 and was thus available to Schuler.

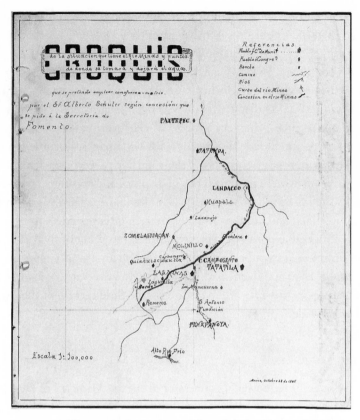

FIGURE 16. The map is the territory: Albert Schuler, *Croquis de la situación que tiene el río Minas,* 1905. Courtesy of the Archivo Histórico del Agua, Mexico City. Photograph by Carmen H. Piña.

Schuler's cartographic attachment warrants further consideration. On the one hand, its inclusion is entirely understandable. The decree of 1894 required solicitants to provide a map or some kind of descriptive report of the region in question. Moreover, given the federal government's own lack of geographic knowledge, particularly of a fairly remote region such as the canyon of Zomelahuacán, Schuler's image could only help push his petition forward rapidly, even more so given that the image had been taken from a map of the federal CGE. The image would effectively, or so Schuler hoped, trump any of those nagging questions that had arisen in the past regarding the names and locations of waterways. After all, how could the government argue against the results and authority of its own surveys? With a petition "satisfactorily conform[ing] with the dispositions of the Decree of 15 June, 1894" and an up-to-date cartographic portrayal of the region, Schuler could reasonably look forward to a quick resolution to his petition.[34]

It was not to be, for reasons that compel an examination of the more complex history behind Schuler's cartographic claims. Three years earlier, in 1902, Schuler had petitioned for rights to the Las Animas River (also known as the Romeros or Trinidad River). The Ministry of Communications and Public Works, in consultation with the Ministry of Development, had refused to act upon his request, having determined that the river was under local rather than federal jurisdiction. Indeed, they deemed *all* rivers in the canyon *other* than the Minas River (and its uppermost portion, known as the Tenepanoya) to be under local jurisdiction. As such, a concession to the Las Animas River could only be granted by the appropriate local body—in this instance, the municipal council of Las Minas.

In 1904 Schuler thus petitioned the municipal council. The members promptly rejected him.[35] The council members feared that Schuler had more interest in speculation than production. The Mine Law of June 4, 1892, had effectively turned mining *concessions* into property *rights* by decreeing that no longer did one have to work a mine in order to retain title to it. A mine title could be revoked only for nonpayment of taxes, effectively promoting large-scale speculation.[36] Speculation was the bane of the municipality, whose inhabitants depended upon work in the mines for their livelihood. The small houses of the mine laborers who had been coming to Las Minas since the mid-eighteenth century dotted the banks of the rivers and the families' attempts to eke out an existence on the sparse patches of available arable land were a daily reminder of the human cost of abandoned plants and mills.[37] Indeed, experience had

34 Schuler to Ministro de Fomento, October 29, 1905, AHA, exp. 57063, f. 3r–v.

35 This petition was pending as of June 22, 1904. Schuler to Ministro de Fomento, June 22, 1904, AHA, exp. 59912.

36 Bernstein, *The Mexican Mining Industry*, chap. 4.

37 Although small communities—Tatatila and Las Minas—took root, most of the valley's small population lived dispersed along the rivers in the crooks and pits of the mountains. Of the municipio's 1,896 counted people in 1888, only 361 lived in the two-square-block *cabecera* of Las Minas. The rest lived in rancherías scattered in the various canyons and along waterways. See "Sección de Catastro: Datos rendidos"; División territorial del Cantón de Jalacingo formado con arreglo a las prevenciones que contiene la Circular No. 9 del Superior Gobierno del Estado, ex-

borne out the municipal council member's concerns. Previous claimants too often did too little to repair the dilapidated haciendas de beneficio receding into the foliage.[38] The building or reparation of such structures had traditionally been the key to the granting of usufruct rights to waterways in Las Minas: "It has been established custom since remote times to grant to each individual who constructed a hacienda de beneficio the use and enjoyment of the waters that cross this Municipio. This was the custom that the old ayuntamientos followed because this settlement, made up entirely of miners, neither had nor has other means of support than mining."[39]

The granting of usufruct rights, overseen by the council, derived from the investiture in the future of the community. That is, it depended on active, rather than paper, *deeds*. Schuler and his partner, however, had made no effort over the previous years to repair or rebuild the San Anselmo and San Francisco haciendas, an idleness that raised the specter of speculation in the council members' minds. Schuler's sole interest, they insinuated, was *self*-interest.[40] The question of control over the waterways thus revolved not only around issues of rights and custom but also around the need to confront perceived threats to the community.[41]

Significantly, the council framed the defense of these rights in nationalist terms. When Schuler argued to the Ministry of Development that local control over the rivers hindered national progress, suggesting that "the Ayuntamientos wish to do nothing but cause problems and lay

---

pedida el 15 de febrero de 1895, ACGE, exp. 7; and "Memoria presentada al H. Congreso del Estado de Veracruz Llave, por su Gobernador Constitucional el C. Francisco Hernández y Hernández, el día 13 de Marzo de 1869," in Blázquez Domínguez, ed., *Estado de Veracruz*, 2:701.

38 See, for example, the comments in the certificate signed by José Francisco Salas G., dated April 3, 1887, included in José Francisco Salas García, solicita confirmación de sus derechos al uso de las aguas del río La Trinidad, September 12, 1908, AHA, exp. 16333.

39 "Sesión ordinaria del día quince de Junio de mil novecientos diez," included in José Francisco Salas García to Secretario de Fomento, August 31, 1910, AHA, exp. 16333.

40 See, for example, Alcalde Municipal to Jefe Político del Cantón de Jalacingo, October 30, 1906, AHA, exp. 57063, ff. 62–64.

41 For a similar argument applied to a dispute in neighboring Puebla, see Aboites and Morales Cosme, "Amecameca, 1922."

traps, thereby impeding the progress of the exploitation of the mines," the president of the ayuntamiento responded with a sharp rebuttal in a letter of his own: "[T]he president of the Ayuntamiento would like to inform you that he is perfectly willing to consider solicitudes based on *good faith and concerned with promoting industry and contributing to the progress of the nation.*"[42]

Members of the council were not simply protectors of the populace. They had their own political and economic interests to tend to. A variety of powerful individuals had usufruct rights to the waterways, and certainly Schuler's operations would have impacted them. Indeed, council members offered the above-cited explanation of historical use rights on behalf of José Francisco Salas García, a powerful individual who had been involved in local politics over the years and who was eager to protect his own mining operations and water concessions.[43] (After the revolution, Salas García would in fact lead the opposition against villagers' attempts to receive a grant of ejido land in the valley.)[44] Moreover, Schuler petitioned for a set amount of water over an indefinite period of time—ten thousand liters per second—regardless of the season. His desire for a fixed grant of indefinite duration to a very "unfixable" resource—in effect, a property right rather than usufruct claim—could spell disaster for other users.[45] No wonder members of the ayuntamiento wanted to maintain the power to determine how the waterways

42 Alberto Schuler and E. Guillermo Vogel to Ministro de Fomento, September 29, 1906, AHA, exp. 57063, f. 59; Alcalde Municipal de Las Minas to Jefe Político del Cantón de Jalacingo, transcribed in Jefe Político de Jalacingo to Secretario de Estado y del Despacho de Fomento, Colonización, e Industria, October 30, 1906, AHA, exp. 57063, ff. 62–64. My emphasis.

43 Salas García's position in the community is suggested in part by his signature on reports to the prefect. See, for example, "Sección de Catastro: Datos rendidos."

44 See AGEV, CAM, exp. 1121, Municipio de Las Minas, Poblado Las Minas, Dotación de ejidos.

45 Duration, in Carol Rose's terms, serves as the "crucial component which differentiates between property and mere usufruct. When one claims a right to something even as changeable and literally fluid as water, one may take water from the stream here and now; but in an appropriative system, what makes this act a property claim is the potential right to take the same amount every year, into the indefinite future. To be sure, some property claims, like leases, may be bounded in time by their very terms, but even those claims are supposed to stay the same over the duration of the specified entitlement." Rose, *Property and Persuasion*, 272.

would be used, by whom and at what cost. Finally, their opposition also stemmed from the fact that they were simultaneously engaged in negotiations with the Compañía de Altos Hornos to sign a profitable contract for the use of the river's waters, under the terms of which the company would pay a rental fee amounting to 720 pesos per year.[46]

The ayuntamiento's fears and interests condemned Schuler's petition to a quick death. What was Schuler to do? His dreams were drowning in two small waterways in a remote canyon in eastern Mexico. He had purchased a substantial amount of land and old mills but neither were of much value without the use of the waterways. With such heavy investments, he could not simply let the matter be, as if with only one failed effort the process had run its course. He responded by attempting to circumvent the municipal council by petitioning directly to the federal government in 1905. Such a strategy required that he convince the staff in the Ministry of Development that the river in question—the tributary known variably as the Las Animas, Romeros, or Trinidad River—fell under *federal* jurisdiction. If the Minas River had been definitively determined to be under federal jurisdiction, his quest now was to convince the government that what the locals called Trinidad or Romeros was actually the Minas. And this is the context within which to understand Schuler's inclusion of his map.

A closer examination of Schuler's map reveals why. Schuler petitioned for rights to the Minas River, and he claimed to clarify his concession by including this map. However, what Schuler traced as the requested concession on his map was a portion of the Minas *as well as* the full length of the Las Animas (Trinidad or Romeros) tributary. The visual effect of the image, with the heavy line marking the requested concession, was to suggest that this tributary was in fact a *continuation* of the Minas, and therefore, already under federal jurisdiction. The effect is all the more powerful for the purposeful silences on the image. The only area where Schuler actually labeled the river as the Minas River is downstream (north) of the intersection with the tributary. Both the uppermost portion of the Minas (the Tenepanoya) and the entire length of the

46 Angel Zavalza to Secretario de Fomento, August 20, 1907, AHA, exp. 60048. Altos Hornos also mined iron in a neighboring municipality and was building a cable line from the valley floor to the railway line at Las Vigas. See Capilla, "Los yacimientos de fierro de Tatatila."

tributary remain unnamed. In effect, Schuler renamed—or *unnamed*—the waterways in such a way as to allow him to petition the federal government for a concession to the Minas yet still garner rights to the tributary Las Animas. Schuler's map brought the landscape in line with his desires.

The Ministry of Development did not examine the map in great detail, nor did it compare it with others available. If it had, it might have predicted the forthcoming storm of protest. Instead, it processed his petition and, in accordance with federal law, published it in the *Diario Oficial*, where it attracted the attention of the ayuntamiento. Although his map was not reproduced, the words were clear enough for the council: Schuler had petitioned the federal government for rights to a river under local jurisdiction. The council members immediately submitted their protest to the ministry—a long document supplemented by numerous attachments from the municipal archive.[47] They argued their case on multiple grounds. In the first place, they stated that Schuler was doing nothing more than trying to bypass their authority. His 1904 petition to the ayuntamiento, combined with his years of residency in the canyon, "plainly proved that the claimant recognized the rights of the Ayuntamiento of Las Minas" to the river in question.[48] Such recognition was further demonstrated, they claimed, by the fact that Schuler never protested their rejection of his petition.[49]

The municipal council then exposed Schuler's deliberate obfuscation. Although he claimed to be petitioning for a concession to the Minas River, in actuality, they argued, he had delineated the trajectory of the Romeros (Trinidad or Las Animas) River. This was the very same waterway he had previously, and unsuccessfully, sought to acquire! To assure the ministry of the truth of their claim, the council members offered a locally grounded explanation: the name of the river, they wrote, "is taken from the settlement in which it originates [Romeros], a part of this municipality, and [is] known also as the 'Río de la Trinidad' as it passes beneath an old bridge with this name."[50] Documents collected by

47 Alcalde Municipal de Las Minas to Jefe Político de Jalacingo, transcribed in Jefe Político to Secretaría de Fomento, January 3, 1906, AHA, exp. 57063, ff. 32–34.

48 Ibid., f. 32v.

49 Ibid.

50 Ibid., f. 32r.

the prefect in previous years, and subsequently submitted to the CGE, concurred.[51] And this river, the council members stressed, had been determined by the federal government to be under local jurisdiction, a decision recently reaffirmed by the secretary of communications and public works. The council forwarded their opposition, as well as all reference material, to Governor Dehesa, who passed it along to the Ministry of Development. By early 1906 Schuler's request had been put on indefinite hold.

The council's methodical and well-documented opposition to Schuler's claim led the two ministries involved in water concessions—the Ministry of Development and the Ministry of Communications and Public Works—to investigate the history of water rights in the region. The official assigned the task found himself having to navigate a profusion of competing claims and names as labyrinthine as the complex array of waterways in the canyon. His research took him through the previous decade of claims and counterclaims by U.S., German, French, and Mexican investors, most of whom still had pending petitions that further complicated and confused the picture of use rights and the names of rivers in the canyon.

He completed his report in early February 1906.[52] He offered two primary conclusions: first, that the Minas River was indeed under federal jurisdiction and that the river variably known as the Las Animas, Trinidad, or Romeros River was under local jurisdiction. However, he reported that he could *not* verify with certainty which of these waterways was which. The multiple names and the confusing web of waterways and tributaries inhibited the resolution of the most basic question of all: which waterway *was* the Minas and which was the tributary. Thus his second conclusion that the conflict would not be resolved until some kind of definitive and fixed representation of the hydrography of the canyon had been established.[53]

51 See "Sección de Catastro: Datos rendidos."

52 See the report of the Ministro de Fomento, February 3, 1906, AHA, exp. 57063, ff. 39–40.

53 The summation of the case is by one Sr. Canaliza, February 3, 1906, AHA, exp. 57063, ff. 39–40. The original determination regarding the Minas River was made June 30, 1897, in response to the petition of Carlos Rivas. Agustín del Río to Secretaría de Fomento, May 1, 1897, AHA, exp. 60085; Carlos Rivas to Secretaría

He forwarded his summation to the secretary of communications and public works who, one month later, made a new decision designed to clear up the confusing panorama.[54] He first reaffirmed the status of the various rivers before then offering his own cartographic clarification of their names and locations: a map "in which has been marked in red the Minas River and its upper course named Tenepanoya, both declared to be of federal jurisdiction; and indicating in yellow the course of the rivers Frío, Puerco, Tequesquite, San Francisco or Zomelahuacán, and Trinidad or Romeros or Lagunilla or de las Animas, which are under local jurisdiction" (figure 17).[55]

Here then, was the final word on the names and locations of the myriad rivers. Or was it? Exactly what sources the ministry relied upon in creating the map is unclear. What is certain, however, is that, although intended to reconcile a confusing and contradictory landscape, the image did not have the desired effect. In fact, the map generated more problems than it solved. When copies of the verdict and map were dispatched to Schuler and the council, both parties raised their voices in protest.

The council viewed the map with ambivalence. It noted its gratitude to the secretary for reaffirming that the Romeros River, among others, was indeed under local jurisdiction, but it also expressed concern that the Minas had been determined a federal waterway. The council claimed that it had no knowledge of the resolutions from 1897 declaring certain rivers under local control and others under federal jurisdiction. Moreover, it noted,

---

de Fomento, May 18, 1897, AHA, exp. 60085. The determination of the status of the Las Animas, Trinidad, or Romeros River was made in response to the series of solicitations made by Schuler and the Vogel family on February 28, 1902. See Schuler to Ministro de Fomento, February 28, 1902, AHA, exp. 60104, and Vogel to Secretario de Fomento, February 28, 1902, AHA, exp. 57049. For the official determination, see Secretaría de Communicaciones to Secretaría de Fomento, March 14, 1902, AHA, exp. 60104, and Secretaría de Communicaciones y Obras Públicas to Secretaría de Fomento, March 14, 1902, AHA, exp. 57049. The same verdict was reiterated a year later; see Sección Primera de Comunicaciones y Obras Públicas al Secretaría de Fomento, April 27, 1903, AHA, exp. 60301.

54 Secretaría de Comunicaciones y Obras Públicas to an unknown recipient [incomplete folio], date unknown but mid-March, 1906, AHA, exp. 57063, f. 47.

55 Ibid.

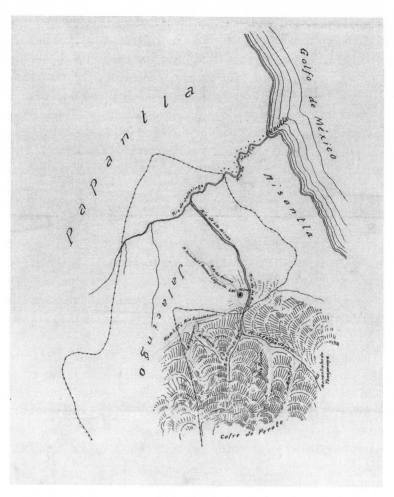

FIGURE 17. Is the map the territory? Secretaría de Comunicaciones y Obras Públicas, untitled map of the Minas River and its tributaries, 1906. Courtesy of the Archivo Histórico del Agua, Mexico City. Photograph by Carmen H. Piña.

this map marks rivers that do not exist in such places, it situates them in areas through which only the Puerco River flows, and it alters the location of the Romeros River; this Commission does not understand the reason as to why the author of this map has made those alterations. . . . [W]hat is termed Tequexquite, or Miqueta, San Francisco or Zomelahuacán is not a river, and not even in the position in which it has been put.[56]

In effect, the state map proved locally unintelligible and reflected little of the reality in which the inhabitants of Las Minas resided.

For Schuler, the map held more ominous portents. The ministry's decision declared the Minas to be under federal jurisdiction, but its map deemed the portion of the river desired by Schuler to be the Romeros and thus under local jurisdiction. It was now May of 1906. After more than six months of debate and contestation, Schuler found himself back where he started. For both Schuler and the ayuntamiento, questions and concerns persisted, and the paper landscape left much to be desired. As the dispute unfolded, both Schuler and the ayuntamiento again sought refuge in cartographic arguments to promote their version of the landscape.

III. COMPETING CARTOGRAPHIES

Schuler simmered over the continual complications and spent the summer cooling off in the rains of Mexico City before going back on the offensive in September 1906. While time seemed to be slipping through his fingers—it had now been over four years since his first request for rights to the river—all was not a loss. After all, he now had a version of the state's conception of the landscape in hand, an imprinted image to which he could respond and around which he could organize his claim. No longer having to refute the (in his mind) infuriatingly confused assertions of the ayuntamiento, he now devoted his attention to convincing the federal government of the basis of his petition.

Schuler had not been idle over the summer of 1906. He spent the latter part of July and August awaiting the results of a hydrographic survey

56 Alcalde Municipal de Las Minas to Jefe Político del Cantón de Jalacingo, November 12, 1906, AHA, exp. 57063, f. 62v.

being conducted in the canyon by two engineers he had hired. By September they had finished their work and submitted a report. Schuler immediately contacted the Ministry of Development:

> [W]ith respect to some of these rivers there still exists some uncertainty as to whether or not they are under federal jurisdiction, owing to the fact that antique maps were more or less incorrect and owing also to the fact that the inhabitants of the villages along the rivers are in the habit of giving a single river various names. . . . [Thus] we submit a map . . . constructed by two competent Engineers who we have had here for more than two months, at considerable expense, to make a close study of the region and its rivers, and whose map . . . is in agreement with that of the Geographic Exploration Commission.[57]

Once again Schuler used the CGE map as the basis for his argument of how the landscape should in fact be construed. In other words, he offered, at his own expense, a confirmation of the state's own independently constituted facts and a refutation of the image he had received from the secretary of communications and public works.

Schuler, however, realized he could not simply attempt to trump the most recent cartographic image of the region. He had already argued his case in this way and had come up short. Thus, after meticulously elaborating upon the construction of this new map and its concordance with that of the CGE, Schuler made an interesting statement: "It is the case that even assuming, although not conceding, that the Animas (or Trinidad etc.) River were not the same river as the Minas, not only is it navigable but it contains nearly double the volume of water of the Tenespanoya [*sic*]."[58] The Tenepanoya, it will be recalled, had been deemed the upper portion of the Minas and thus under federal jurisdiction. At the moment of making his most comprehensive explanation, he wavered, hinting that the Minas River may not have been the same as the tributary to which he sought rights. Why? Most likely, Schuler recognized that his best bet for garnering rights to the river now lay in emphasizing those conditions he believed would effectively make a river federal—being a boundary with another state, navigable, or containing a volume of water substan-

57 Alberto Schuler and E. Guillermo Vogel to Secretaría de Fomento, September 29, 1906, AHA, exp. 57063, f. 59r.
58 Ibid.

tial enough to be considered a national resource. The first criterion was impossible to argue, but the latter two were not, particularly considering the state's persisting paucity of information.

This was not the first time such an argument had been made. Schuler's partner, Guillermo Vogel, used a similar line of reasoning in 1903. In a request for rights to the waters of the Las Animas, he included a small map, noting that "being that the Río de Las Animas, although neither navigable nor *flotable*, is the largest tributary of the Minas and this being a tributary of the Nautla, as you can see in the attached sketch, it should in my opinion be considered under federal jurisdiction."[59] Vogel's attempt did not succeed, but this did not dissuade Schuler, who proffered the same argument with some extra embellishment: he suggested that the Las Animas *was* navigable and then provided further cartographic amplification. In contrast, he suggested, the Tenepanoya was neither navigable nor did it contain a significant volume of water. Thus the logical continuation of the Minas should be the Las Animas rather than the Tenepanoya. He made a normative claim based primarily upon water volume and hence resource value. It was illogical that a waterway with a smaller volume of water should be a continuation of a given river, while a waterway with a larger volume should be condemned to tributary status.

Schuler's perspective regarding how use rights and jurisdiction should be understood varied radically from that of the municipal council. Like Schuler, the council submitted a map. However, in sharp contrast, it was an ambulatory, narrative map: "[T]he Minas cannot be considered a federal river, since it consists of the small arroyo of Tenepanoya, of the arroyo called Río Frio, and the arroyo named Río Puerco, until they join at the foot of the hill named Nopalteptl from where it takes that name because it encircles this town to the south and to the east and then joins its waters with that of the Romeros to the northeast of the same town."[60]

This map bore an obvious similarity to the kinds of narratives pro-

59 Vogel to Ministro de Fomento, March 10, 1903, AHA, exp. 60301. The negative opinion of the minister of communications and public works was issued the following month. See Sección Primera de Comunicaciones y Obras Públicas to Secretaría de Fomento, April 27, 1903, AHA, exp. 60301.

60 Alcalde Municipal de Las Minas to Jefe Político del Cantón de Jalacingo, November 12, 1906, AHA, exp. 57063, f. 62v–63r.

duced during the perambulations of village or parish boundaries. Michel de Certeau has called such perambulatory narratives "tours," in contradistinction to "maps."[61] His distinction is designed to not only differentiate between different forms of spatial representation; it is also meant as a consideration of the spatial logic of power. For de Certeau, the logic of power is inevitably wrapped up in the *map*, its godlike perspective reflective of an epistemological and political position. In contrast, the *tour* constitutes the code of the user, a contextual and horizontal perspective derived from daily practices. De Certeau's distinction is too dichotomous, but it is nevertheless useful in considering the differences between Schuler's submissions and those of the town council. In the latter's vernacular map, or tour, the waterways were not fitted into a larger hydraulic system based upon their relative size, volume, or commercial potential; there was no obsession with tracing the river to a "source"; no parenthetical denoting that the Tenepanoya constituted the "upper portion" of the Minas; no notion of a river as a principal channel, fixed and linear. Rather, the narrative emphasized the historical and geographical specificity of the rivers in question; it implicitly distinguished between rivers and arroyos, reflecting understandings of seasonal variation; and it connected names and passages to their human context and history. The Minas River appeared not as a linear, fixed passage traced on a map, but as the confluence of multiple arroyos.

Yet the council members understood that their tour could hardly stand alone as their only piece of evidence. To make their claim audible, they had to engage the developing state to some degree on its own terms, with its own idioms: through the objective map. In effect, they had to update their discourse. Thus, they referred the ministry to another federal map; specifically, sheet 19 of the Army and Navy Geographic Commission's 1894 General Map of the Mexican Republic. "Although there are some differences regarding the area," they wrote, "it places the Minas River in its natural position."[62] Their reference serves to underline just how confusing and contentious the entire process of standardization could be: they used one federal map (that of the Army and Navy Geographic Commission) to counter another federal map (that of the Geographic Exploration Commission).

---

61 See de Certeau, *The Practice of Everyday Life*, 118–21.
62 I have been unable to locate a copy of the map so far.

For all intents and purposes, the conflict had now degenerated into one of competing cartographies. Although assuredly more complex than that, it is worth observing how much the resolution of the conflict did indeed depend upon a cartographic determination, a validation of one representation over another. In such circumstances, the map was not, as Denis Wood has observed, "an innocent witness . . . but a committed participant, as often as not driving the very acts of identifying and naming, bounding and inventorying it pretends to do no more than observe."[63]

In late 1906 the secretary of communications and public works, in response to the previous two letters, commissioned an engineer to reconnoiter the region and make a final decision.[64] The engineer visited the region, surveyed the rivers, and compiled an enormous amount of data on water flow, preexisting rights, and existing water works. He submitted his report and, now eight months later, the secretary of communications and public works reached a decision. Schuler finally got the news he wanted. The secretary reaffirmed the Minas to be under federal jurisdiction and that the Tenepanoya constituted its upper portion. At the same time, he determined the Trinidad (or Romeros or Las Animas) to be under federal jurisdiction because "it is an important tributary of the Minas."[65] The ministry effectively confirmed the council's representation of the rivers but then federalized them all. Over the coming years, the name of the tributary would slowly be ground down to a single name: Trinidad.

Ironically, Schuler never received his petition. On December 17, 1908, the Ministry of Development signed a contract with Angel Zavalza, a representative of the Compañía de Altos Hornos Mexicanos, "for the use, for power generation, of the waters of the Rivers Minas and Trinidad."[66] The ministry, having appropriated from the ayuntamiento

63 Wood, *The Power of Maps*, 79.

64 Secretaría de Comunicaciones y Obras Públicas to Secretario de Fomento, December 8, 1906, AHA, exp. 57063, f. 74.

65 Secretaría de Comunicaciones y Obras Públicas to Secretario de Fomento, August 5, 1907, AHA, exp. 57063, f. 80.

66 "Contrato celebrado con el Sr. Angel Zavalza, en representación de la Compañía Altos Hornos Mexicanos, para el aprovechamiento, como fuerza motriz, de las aguas de los ríos Minas y La Trinidad, del Estado de Veracruz," *Memoria presentada al Congreso de la Unión . . . 1908–1909*, li.

the power to confirm contracts and future rights over the region's waterways, had given something back: it had respected the wishes of the ayuntamiento by confirming their contract with Altos Hornos.

CONCLUSION

The conflict did not simply end. Schuler persisted in his efforts to procure rights to the waterways until his violent death, presumably at the hands of bandits, on the road connecting Xalapa to the port of Veracruz, four days before Christmas in 1916.[67] What did end, however, was the "confusion" over the rivers Minas, Romeros, Trinidad, and Tenepanoya, each of which, by the end of the revolution, had attained a documentary permanence, fixed on more parchment in a growing state archive.[68] Yet *how* those names became fixed is as interesting and important as the mere fact that they were fixed. If the mass of contradictory and confusing documentation the case generated is, on the one hand, the most compelling testimony available of the very need to fix names and waterways, it also reveals just how contested the process of producing those fixations could be.

Maps, and their content, are often assumed to be the product of the hard and risky work of engineers and surveyors, of reluctantly supportive state bureaucracies, and of eager scientific societies. While the specifics may change, the protagonists remain the same. Certainly their role was fundamental. But the spatial construction of Mexico, its mapping and naming, was not an exercise solely under their purview. It was just as frequently a product of struggles over maps, documents, titles, and names between and among various groups: town councils, villagers, hacendados, mine owners, entrepreneurs, and government officials. What appeared on a map resulted as much from local conflicts and negotiation as it did from a presumably objective, instrument-based survey. In the case of Las Minas the state had an array of purportedly definitive

67 Gabriel Ortíz González al Secretario de Fomento, AHA, exp. 60048, f. 201r–v.

68 Para que se publique en la Gaceta Oficial del Estado, remite copia de la Ampliación de la Declaración de Propiedad Nacional de las aguas del río Trinidad, AGEV, RF, Sección de Aguas, Las Minas, Manantiales: año 1927, num. 1, II.424(87); Samuel Hernández y F., "Red hidrológica del Edo. de Veracruz," Secretaría de Agricultura y Fomento, Dirección Estudios Geográficos, Depto. Hidrología y Topografía, Sección Hidrología, exp. 17, ACGE, carpeta 125.

maps available for consultation. The CGE had not only completed its traverse surveys of the region, the maps had been plotted and published. If they were readily available to Albert Schuler, certainly the federal government could not claim ignorance. Yet the CGE map deployed by Schuler, surprisingly enough, did not emphatically put to rest any further questions regarding the location and names of the various waterways. CGE maps may have been designed to constitute definitive state representations of the land, but in the context of highly charged encounters among communities, capitalists, and the state they rarely held such sway. As William Roseberry has observed, while the "state claims the power to name; to create and print maps with state-sanctioned labels," that does not mean such images and names become the de facto currency of daily life.[69]

In the process of attempting to construct and organize its knowledge of the provinces, the state relied upon numerous, competing voices and visions. It was by assuming the right to mediate and arbitrate such disputes that it simultaneously legitimated itself *and* acquired the kind of knowledge necessary to administer and rule more efficiently. The state did not necessarily impose itself upon local populations like some modernist juggernaut; rather, it assumed power and legitimacy by mediating conflicts—conflicts generated by its own legislation. In this instance, the federal government, with one hand, confirmed the ayuntamiento's contract with Altos Hornos, and with the other, appropriated power over the rivers. The last piece of documentation from the Schuler file punctuates with finality this new relationship: "the Alcalde Municipal of Las Minas has been ordered to make known to the users of the waters of the Río Minas that they do not possess titles that authorize their rights to these waters."[70] This applied equally to Schuler and to Salas García, to Vogel and to the villagers. Titles, granted federally through a process of application and bureaucratic approval now determined rights to the waterways. The long-enjoyed and entrenched local rights of ayuntamientos were not extinguished like the flame of a candle: they evaporated, gradually, like water in the heat of the Veracruz sun.

69 Roseberry, "Hegemony and the Language of Contention," 361.

70 Gobierno del Estado de Veracruz, Departamento de Estadística, Sección de Fomento, to Secretario de Estado y del Despacho de Fomento, Colonización e Industria, October 24, 1908, AHA, exp. 57063, f. 136.

CHAPTER 7

Revolutionary Spaces

In a prescient piece of writing, Andrés Molina Enríquez, lawyer and honorary member of the Sociedad Mexicana de Geografía y Estadística, wrote that "the revolution in France not only disentailed the lands of the clergy but also of the nobility. We desire a similar achievement. . . . [I]t is necessary and it will be done, either through the peaceful means which we suggest or by a revolution which sooner or later must come."[1] The year was 1909, and the revolution came sooner rather than later. Eventually, Mexico's nobility—the infamous class of hacendados—saw much of their land redistributed to agricultural communities in the form of ejidos (inalienable concessions of land granted to communities by the state).

All times have their spaces through which ideologies are inscribed, codes embodied, histories redefined, and social relations reconfigured.[2] Postrevolutionary Mexico's was the ejido. The ejido functioned as a primary mechanism for returning land to the pueblos, redistributing it to the landless and fulfilling the promise of the revolution. It also constituted the means by which to create new relationships of reciprocity and dependency between the state and rural cultivators. At the same time, it functioned as the spatial prism through which an entire corpus of revolutionary history (and mythology) was refracted. With its nominal linkages to both the 1855 Liberal Revolution and the Aztec calpullalli, a new state construct appeared as a return to a number of selected traditions.[3]

1 Molina Enríquez, Los grandes problemas nacionales, 199.

2 Lefebvre, The Production of Space, 44.

3 Numerous authors have noted these linkages. For example, see Mallon, "Reflections on the Ruins"; Mallon, Peasant and Nation; McBride, The Land Systems of

By 1940 at the end of President Lázaro Cárdenas's six-year tenure, the ejido had become a familiar figure across the Mexican landscape. Over the remaining half-century, it became the ideological and material space—the metaspace—upon which the postrevolutionary regime built its foundation.

But consequences are not intentions. Exactly what the revolution and the agrarian reform meant in the waning years of the fighting, and well into the 1920s, was subject to broad interpretation and understanding. Although various factions had begun already to interpret (and reinterpret) the uprising, the revolution had hardly acquired status as *The Revolution*, codified in a closed system of controlled meaning.[4] Rival elite factions, agrarian bureaucrats, hacendados, and villagers: all interpreted revolutionary experience, discourse, and decrees through their own lens of expectations, hopes, and fears, and the manner in which agrarian reform proceeded resulted from struggles and arguments between and among these various actors. In the 1910s and 1920s agrarian reform remained a work in progress, openly contested by a wide variety of actors. This chapter, in tracking the convoluted process of reform during the waning years of armed conflict and through the early years of reconstruction, demonstrates that agrarian reform—indeed, the ejido itself—was neither a fixed idea nor a fait accompli.

I. THE SPIRIT OF THE REVOLUTION

On January 6, 1915, Venustiano Carranza, self-proclaimed First Chief of the Constitutionalist Army, promulgated a new agrarian reform law.[5] Authored by Luis Cabrera, intellectual head of Carranza's forces, the

---

Mexico; Knight, "Racism, Revolution, and Indigenismo"; and Nugent and Alonso, "Multiple Selective Traditions."

4 A useful examination of how various elite factions constructed and reconstructed the revolution can be found in Benjamin, *La Revolución*, part 1. Already in 1911 state governments organized celebrations of "la revolución libertadora," to be held on November 20. See Decreto no. 40 de 17 de noviembre del corriente año, que declara de fiesta en todo el Estado, el próximo día 20 del mismo mes, primer aniversario de la revolución libertadora, AMM, Caja 1911, exp. 21.

5 *Ley de 6 de enero de 1915, que declara nulas todas las enajenaciones de tierras, aguas y montes pertenecientes a los pueblos, otorgadas en contravención a lo dispuesto en la Ley de 25 de junio de 1856*, CVC, fondo 21, carpeta 25, leg. 2477.

law recognized villagers' rights to hold land collectively in the form of ejidos. It thus reversed the liberal model of individual landholding decreed by the Constitution of 1857, which Cabrera viewed as too sharp a departure from pueblo tradition, the cause of villagers' dependent status and, ultimately, of the revolution itself.[6] The new law mandated that lands illegally usurped from communities since 1856 and the passage of the Liberal Reform laws be returned (a process known as *restitución*); meanwhile, pueblos and villagers unable to prove usurpation but effectively landless could ask for a *dotación*—an inalienable grant of land owned by the state but worked in usufruct by the recipients. In order to create such ejidos, land would be confiscated from large haciendas and ranchos.

The articles of the decree superseded existing federal and state legislation and endowed any future federal executive with legal and administrative control over agrarian issues.[7] The modified version of the decree that would be incorporated into the Constitution of 1917 required states to close their agrarian affairs offices and to inform their officials that they no longer had authority to intervene in agrarian affairs.[8] In their place, the federal law generated a vast and centralized agrarian bureaucracy under the aegis of the federal executive; the linchpins of this bureaucracy would be the National Agrarian Commission [CNA] and its regional branches, the local agrarian commissions [CLAS].

The generation of a large legal and bureaucratic structure concerned with agrarian reform did not necessarily mean that reform proceeded apace. Carranza, a large landowner, favored reform out of political expediency rather than ideological principle. In the midst of a power struggle with the forces of Emiliano Zapata and Francisco "Pancho" Villa, he

6 See Henderson, *The Worm in the Wheat*, 54–57. Cabrera's ideas, which built upon those of Molina Enríquez, are best summarized in his *La reconstitución de los ejidos de los pueblos*. Ideas on agrarian justice and agricultural development were not merely the purview of a select few: they were of significant interest early on to officials in both the Comisión Nacional Agraria and the Comisión Agraria Ejecutiva. See Zuleta, "La invención de una agricultura próspera," 265–67.

7 See Gómez, *Historia de la Comisión Nacional Agraria*, 71, 98–101.

8 McBride, *Land Systems of Mexico*, 167; and Gómez, *Historia de la Comisión Nacional Agraria*, 98. Centralization persisted as time passed. For example, in 1926 a new Ejidal Law of Patrimony replaced existing local administrative committees with federal ejidal commissariats that came under the direct control of the Federal Department of Agriculture. See Fowler-Salamini, "Tamaulipas," esp. 209.

needed to create peace and win converts to his cause. As his future son-in-law, Cándido Aguilar, military governor of Veracruz, informed him in 1915, the only way to get rebels in Acayucán, Veracruz to "exchange the rifle for the axe and the plow" was to give them the lands they wanted.[9] Only this would convince them to put down their weapons and turn away from a "revolution [that] has given them absolute liberty," he remarked.[10] Carranza conceded, and agrarian officials worked to resolve the case quickly. But Acayucán is exemplary of Carrancista conservatism: it was one of a small number of highly selective interventions, undertaken "not solely to benefit the petitioners, but because of its great political importance and the need to pacify these regions."[11]

The very structure of the reform process reflected Carranza's tepid feelings about reform. Underneath the progressive patina, Carranza's proclamation elided practical issues of implementation to the degree that one CNA official recalled that he and his colleagues operated "blind."[12] Lengthy delays and "Byzantine" discussions over questions of jurisdiction, the relationship between grants and restitutions, and even the definition of the word "ejido" ensued, leaving CLA and CNA staff with little to show after a year's work.[13] Whether or not Carranza intended to mire the process down before it had even begun, as some have argued, it is clear that he pursued reform not with the passion of a revolutionary but with the strategic calculus of a political pragmatist.[14] As such, he felt little need to refine the legislation in order to facilitate the process.

9 Cándido Aguilar to Venustiano Carranza, February 1, 1915, AGEV, CLA, caja 1, exp. Restitución de ocho sitios de Ganado Mayor que reclaman los naturales de las Congregaciones de Acayucán.

10 Ibid. See also Informe relativo a los derechos que reclaman 27 congregaciones del Cantón de Acayucán sobre 8 sitios de ganado mayor basado en tres expedientes y datos particulares presentados por el Señor José Fernando Ramírez, September 9, 1914, AGEV, CLA, caja 1, exp. Acayucán, 1903–1921.

11 Presidente de la Comisión Local Agraria to Presidente de la Comisión Nacional Agraria, July 24, 1917, AGEV, CLA, caja 1, Acayucán, 1903–1921. Carranza's limp performance is easily measured: from 1915 to 1917 only nine villages in the entire country were granted land through the reform. See Womack, "The Mexican Revolution," 169; Knight, *The Mexican Revolution*, 2:466–69.

12 Gómez, *Historia de la Comisión Nacional Agraria*, 79.

13 Gómez offers numerous, and often humorous, examples of the kinds of discussions that took place in the early years of the CNA. See ibid., esp. 94–98.

14 Simpson, *The Ejido*; Henderson, *The Worm in the Wheat*, 56–57.

Both during and after Carranza's tenure, numerous clarificatory decrees and circulars were issued, but they often proved confusing, if not contradictory. The sheer mass of pronouncements fostered the development of a small industry of explanatory digests designed to help both state officials and campesinos wend their way through the growing bureaucratic labyrinth. Julio Cuadros Caldas's *Catecismo Agrario*, an enormous compendium of agrarian legislation and interpretation first published in 1923, is the definitive example of such texts. More than a simple collection of laws and decrees, the catecismo was a veritable operator's manual for pueblos attempting to recuperate or receive lands, complete with sample letters and forms.[15] Veracruz governor Adalberto Tejeda purchased three hundred copies of the *Catecismo*'s first run and had them distributed to various communities while the CNA also acquired copies for its staff.[16] Even after the publication and distribution of Cuadros Caldas's tome, more clarificatory instructions to both surveyors and villagers appeared regarding as varied activities as the formation of a General Agrarian Census, the creation of maps, proper field methods of survey, and instructions for giving a provisional grant.[17]

Carranza's bureaucratic subordinates did not necessarily share his lukewarm temperament. In point of fact, sharp divisions existed within the agrarian reform apparatus, not least of all between the CNA and the CLA. Carranza created the former, John Womack notes, "to oversee and circumscribe local decisions on villages' claims," as well as the local operations of the CLAS.[18] CLA personnel, meanwhile, performed the most "sensitive and important functions in all cases of restitution or dotación of lands to the pueblos," including determining the size of the grant and

15 Palacios, "Julio Cuadros Caldas," xiii–lvii; "Instrucciones y machotes para pueblos," in Cuadros Caldas, *Catecismo Agrario*, 553–78. See also Villareal Muñoz, *Restitución y dotación de ejidos;* Comisión Nacional Agraria, *Leyes y disposiciones;* and Comisión Nacional Agraria, *Recopilación Agraria.*

16 Palacios, "Julio Cuadros Caldas," xxviii; Gómez, *Historia de la Comisión Nacional Agraria,* 265.

17 See the three documents by Ing. Bartolomé Vargas Lugo, all in AGEV, CLA, caja 4: "Instrucciones para la formación del Censo General Agrario," December 19, 1925; "Instrucciones sobre levantamientos topográficos," December 23, 1925; and "Instrucciones a que deberán sujetarse los ingenieros de la Comisión Nacional Agraria al dar una posesión provisional o definitiva," December 23, 1925.

18 Womack, "The Mexican Revolution," 168.

demarcating it.[19] State governors appointed these officials. Thus, as much as the new legislation centralized power in the federal executive, regional authorities still wielded a significant amount of power over the reform process. While a federal official cautioned governors to "take special care in designating persons who . . . are identified with the generous and patriotic spirit of the Revolution, in order to ensure the most success in the performance of those functions," what constituted success or the spirit of the revolution varied from governor to governor.[20] In many states, gubernatorial opposition to land reform ensured the persistence of old patterns of land holding. The case of Tamaulipas was not unusual: Governor López de Lara, an opponent of land reform, simply dissolved the CLA. When forced to reestablish it, he appointed colleagues and friends who shared his persuasions to serve as its members.[21] Opposition by governors allied with large landowners could be so severe that in 1920 the CNA tried to assume power over CLA appointments.[22] Indeed, CNA surveyor Marte Gómez, who would go on to become governor of Tamaulipas and a confidant of Lázaro Cárdenas, recalled that his agency tended to be much more radical and sincere in its dedication to reform than the CLAS.[23]

Gómez's hindsight was clearly not 20/20. In particular, the four members of the Veracruz CLA in 1918—Nabor Cuervo, Salvador de Gortari, Gonzalo Vázquez Vela, and Rafael Vargas Espinosa—would have questioned his memory. They were in fact repeatedly and bitterly disappointed by both Carranza and the CNA.[24] A survey in Cosamaloa-

19 Pastor Rouaix to Gral. Benjamin Hill, Gobernador del Estado de Sonora, January 19, 1915, CVC, fondo 21, carpeta 25, leg. 2475, doc. 1.

20 Ibid.

21 See Fowler-Salamini, "Tamaulipas," 195–96.

22 Gómez, *Historia de la Comisión Nacional Agraria*, 193.

23 Ibid., 194.

24 Heather Fowler-Salamini demonstrates quite clearly that the CNA dragged its feet, resolving Veracruz petitions at a rate less than half that of the CLA. Fowler-Salamini, *Agrarian Radicalism in Veracruz*, 99. Cuervo had been recommended for a position in the CLA by a friend of Carranza's in 1915. Gabina C. V. de Baillorez to Venustiano Carranza, May 12, 1915, CVC, fondo 21, carpeta 39, leg. 4260. Vargas Espinosa assumed the presidency in 1918, taking over from Manuel Domínguez, on the recommendation of then senator Adalberto Tejeda, who called Vargas Espinosa "a man of eminent revolutionary qualification." Vargas Espinosa had been a surveyor in Chicontepec, the same area where Tejeda himself had worked as a

pan is a case in point. In 1918 CLA surveyor Carlos Rousseau journeyed to Cosamaloapan to survey lands for restitution to the pueblo. On December 20, 1918, he submitted his map and report to the CLA, which soon after approved the restitution. In February, however, CLA officials learned Carranza himself had rejected the survey in order to respect the holdings of the hacienda San Gabriel. Moreover, the CNA sent an angry letter to the CLA ordering them to reverse their provisional grant. The CLA refused, arguing that the actions of the CNA were highly inappropriate. "[W]e work in accordance with the Law," Cuervo stated, "and we shall see who will come out on top."[25] Although they might personally suffer as a result of their principled stance, he continued, "this should not be taken into consideration because we personally mean nothing[;] but if we do not oppose the injustices that they [the CNA] commit, it could rightly be said that we are not serving the purposes to which they have entrusted us."[26] Regardless, the president's decree stood, and Cosamaloapan had to wait more than a decade for further movement on the petition.

Land grants were not the only points of contention. CLA officials often chafed at the CNA's insensitivity to local reality. When CNA officials banned ejido recipients nationwide from destroying monte (wooded) land in order to cultivate, CLA members responded by explaining that two different kinds of monte—*alto* (forest) and *bajo* (scrub)—existed in Veracruz, the latter of which constituted the vast majority of potential ejido land. The CNA's response would be humorous if it were not so tragic: they informed the CLA that all questions related to *bosques* (forests) were to be referred to the Forestry Office.[27]

Debates also erupted within the CLAs—not surprisingly given their eclectic composition: George McBride, during his visit to nine different state agrarian commissions in 1922 and 1923, found "avowed radicals; sane, thoughtful students of conditions in Mexico; impractical theorists; [and] mere office holders, little interested in anything but the sal-

---

surveyor until 1913. See Adalberto Tejeda to Candido Aguilar, October 4, 1918, CVC, fondo 21, carpeta 125, leg. 14094. The context of the appointments of de Gortari and Vázquez Vela is unclear.

25 AGEV, CLA, caja 3, Copias de Actas 1919. Cuervo's remarks are undated but are ca. mid-February 1919.

26 Ibid.

27 Gómez, *Historia de la Comisión Nacional Agraria*, 100.

ary they received," all working on the same commission.[28] The Veracruz CLA was no exception. When landowners in the state, concerned about the effects of an ejidal grant, requested copies of the solicitations or the CLA's maps, members argued over whether or not to provide them. Some, like de Gortari, argued that the maps were made at the government's expense and therefore should not be forwarded to landowners who might use them "against the rights of the solicitors of land" or "for different purposes" than those for which they were intended.[29] Others disagreed. In early 1919 Trinidad Herrera, a large landowner in Huayacocotla, requested a copy of a solicitation for restitution submitted by villagers in the region. De Gortari strongly opposed Herrera's request, suggesting it would prejudice the interests of the campesinos and compromise the work of the CLA. "Genuine revolutionaries," he argued, worked "to protect the weak."[30] But Cuervo voiced a dissident opinion: "The true revolutionaries in this Commission will be those who apply the revolutionary laws that guide us with the serene and impartial criteria of justice and equality."[31] De Gortari viewed such an understanding of "revolutionaries" as too limited. Vargas Espinosa agreed: the criteria for judging what was revolutionary could not be so "legalist," he argued; rather, the point was to not serve "degenerate interests."[32]

Eventually the issue went to a vote, which served only to broaden

28 McBride, *The Land Systems of Mexico*, 161 n.4.

29 AGEV, CLA, caja 3, exp. Copias de actas de la Comisión Local Agraria, Acta 123, January 14, 1919, and Acta 124, n.d. (ca. late January 1919).

30 AGEV, CLA, caja 3, exp. Copias de actas de la Comisión Local Agraria, Acta 126, February 11, 1919. This was not the only time de Gortari voiced such an opinion. Regarding a conflict over water rights between a large company and a community in Orizaba, he argued that it was preferable to act in the best interests of the pueblos, even at the expense of industry, "because to do otherwise is what had caused our revolutions." AGEV, CLA, caja 3, exp. Copias de actas de la Comisión Local Agraria, Acta 125, February 24, 1919.

31 AGEV, CLA, caja 3, exp. Copias de Actas de la Comisión Local Agraria, Acta 126, February 11, 1919. It is unclear why the date for this Acta is prior to that of Acta 125.

32 Ibid. It is worth mentioning here that landowners who faced dispossession often refused to submit their maps when requested to do so. See, for example, Daniel Muñoz Esteban, Agente de Propoganda Agrícola en el Estado, to Presidente de la Comisión Local Agraria, June 12, 1918, in AGEV, CLA, Caja 4 (1922–1929), exp. Misantla.

the rift. Cuervo and a fourth member, future governor of Veracruz and President Lázaro Cárdenas's minister of education Gonzalo Vázquez Vela, both voted to submit the requested materials to Herrera; de Gortari opposed them. At that point the voting should have ended as the president did not have a vote in such affairs unless there was somehow a tie. But instead, Vargas Espinosa cast a "no" vote, effectively *producing* a tie. A heated argument ensued, during which Vargas Espinosa opined that he had followed the spirit, if not the letter, of the Agrarian Code. Defending himself, he argued that Herrera only wanted copies of the petitions in order to delay the process further. The people of Huayacocotla had suffered repeatedly at the hands of Herrera's supporters, he concluded, so why help him now?

Personnel in the agrarian bureaucracy were not passive recipients of the discourses and plans of their superiors. They were agents in their own right who interpreted revolutionary discourse and agrarian legislation through their own lens of expectations and beliefs, and they struggled with their superiors and among themselves over the implementation and course of agrarian reform. Thus it was that large landowners in Veracruz could confront an entirely different (and more ominous) future than their peers in neighboring Tamaulipas. As a consequence, they sought out other nodes of influence. They turned their attentions to the most prominent figure in revolutionary reconstruction and agrarian reform: the surveyor.

II. HONORABLE REVOLUTIONARIES

On a Wednesday morning, in October 1923, José Alvarez awoke in his temporary quarters near the railway station in Omealca, Córdoba, to another day of work. A surveyor in the employ of the Veracruz CLA, he had arrived the previous month to meet with the campesinos and carry out the initial surveys necessary for a provisional land grant. Around nine that morning, his door burst open and an individual, "gun blazing" and weighed down with ammunition belts, entered. The assailant stated that he represented a regional Union of Landowners and Farmers who wanted Alvarez to abandon his work and leave the municipio immediately. Confronting Alvarez in an "aggressive and threatening manner," he claimed that "all the authorities and *agrarias* of the State, including the Governor, were a collection of shameless thieves, while he, and those with whom he worked . . . were a kind of Klux Clan [*sic*] who would descend upon the surveyors when they least expected it."[33]

He departed soon after to join a number of his self-proclaimed "Clan." They paid a visit to a large group of campesinos finishing the roof on a building designated to be the new Agrarian Office. "Don't be such asses," he told them, his pistol drawn. "You're only building a house for a bandit—the Governor of the State—and all the agrarian authorities." The campesinos held their tongues and the Clan rode off.[34]

Alvarez immediately dispatched an urgent telegram to the governor requesting that a military detachment be sent in to put a stop to this kind of threatening behavior, which had kept the petitioners in a permanent state of "profound fear."[35] His request for military support was ironic, as villagers soon identified the perpetrator as Lieutenant Colonel J. Ortíz Cadena, still in active military service. Alvarez appeared ignorant of a basic fact that former governor Cándido Aguilar had understood years earlier: "honorable revolutionaries are few."[36] Indeed, Ortíz Cadena was only one of an increasing number of high-ranking Carrancista officers who themselves had entered the ranks of the landowning oligarchy and were concerned about the pace of agrarian reform under state governor Adalberto Tejeda (served 1920–1924; 1928–1932). The notorious Carrancista general Guadalupe Sánchez, for example, came to the defense of the Veracruz hacendados—among whose numbers he himself could now be counted—by distributing arms and ammunition to the landlords' *guardias blancas* (hired guards).[37]

33 Comité Particular Ejecutivo de Omealca to Gobernador del Estado, October 26, 1923, AGEV, CAM, exp. 247 (Ampliación), Municipio San Antonio Tenejapan, Congregación Omealca, Dotación de ejidos [hereafter exp. Omealca].

34 Ibid.

35 Telegram included in Gobernador Constl. del Estado to General Alvaro Obregón, October 15, 1923, exp. Omealca.

36 Cited in Knight, *The Mexican Revolution*, 2:385. Emiliano Zapata understood this as well. In an open letter of disgust, he decried the rise of a new class of "modern landowners that use epaulets, military caps and a gun in their belts." "The people," he lamented, "are being mocked for their hopes." Quoted in Aguilar Camín and Meyer, *In the Shadow of the Mexican Revolution*, 65.

37 See Fowler-Salamini, *Agrarian Radicalism in Veracruz*, 37; Knight, *The Mexican Revolution*, 2:469; and Gómez, *Historia de la Comisión Nacional Agraria*, 344. On the entrance of revolutionary generals into the landed oligarchy, and their

Events such as these were not anomalies in postrevolutionary Veracruz. Landowners lived nervous lives as the state government attempted
to fulfill promises of land distribution to peasants. Surveyors, perceived
as subversive radicals turning paper promises into topographic truth,
quickly became the primary targets of large landowners attempting to
protect their interests. Landowners, both old and new, were still powerful enough not to have to resort to the kinds of epistolary anonymity
that characterized villagers' attempts to ward off Porfirian surveyors.
Their confrontations were more audacious. Sánchez not only armed
the guardias blancas; he ordered two of his generals to systematically
prevent CLA surveyors from performing surveys.[38] Ortíz Cadena's well-
armed "Klux Clan" gathered in the plaza of Omealca one Sunday to put
out the word that anyone who accompanied Alvarez on his survey would
be severely punished. The campesinos were not cowed. They requested
weapons from the government so that they might accompany Alvarez
on his survey and not "die like lambs" at the hands of a "mob of *gauchu-
pines.*"[39]

Surveyors were not only threatened with grievous bodily harm. In
other instances they were held against their will. Ismael Aviles, a former
lieutenant in Carranza's army assigned to survey lands in Veracruz, arrived in Acayucán in 1922. He soon complained that a local judge, along
with a representative of the Dutch La Corona oil company, John Mac-
Grath, were obstructing his work.[40] The judge owned four thousand
hectares of land in the area, rented out to, among others, MacGrath's
company. MacGrath, meanwhile, appeared insulted that any Mexican
would dare meddle in affairs under his trust. His company, he reportedly
shouted in a drunken rage, "was accustomed to removing and putting in

relationship to the postrevolutionary regimes of Carranza, Obregón, and Calles,
see Córdova, *La ideología de la revolución mexicana*, and Hamilton, *The Limits of
State Autonomy*, 67–103.

38 Fowler-Salamini, *Agrarian Radicalism in Veracruz*, 37.

39 Comité Particular Ejecutivo de Omealca al Gobernador del Estado, October 26, 1923, exp. Omealca. *Gauchupín* is a pejorative word for "Spaniard." The use
of the term was quite common in the 1920s and seems to have been used fairly indiscriminately as a derogatory term not only for Spaniards but also foreigners and
antiagrarista landowners more generally.

40 Ismael Aviles, Ingeniero de delegación de Depto. de Fomento y Agricultura, to Depto. de Fomento y Agricultura, October 6, 1922, AGEV, RF, Tierras, Deslindes.

FIGURE 18. A new day dawns with tripod, chain, and level: Anonymous drawing celebrating the surveyors of the Veracruz Local Agrarian Commission, ca. 1929. Courtesy of the Archivo General del Estado de Veracruz. Photograph by Bulmaro Bazaldúa Baldo.

place Kings in Europe and had enough money to buy two or three Mexican governments."[41] MacGrath and the judge eventually managed to temporarily imprison Aviles on charges of damaging personal property when he cut down brush in order to do his survey.[42]

The fetishization of the surveyor as the agent of revolutionary change is entirely understandable. Postrevolutionary reconstruction was waged as much with measuring chains and levels as it was with firearms. State-employed surveyors were the next wave of revolutionaries who, armed with tripod, chain, and level, saw themselves as helping a new day dawn across the Veracruz countryside (figure 18). The state surveyor constituted the crucial pivot point between a revolutionary regime and rural society during reconstruction. As one participant put it, they crisscrossed the landscape in order to "enact agrarian justice . . . orient the campesinos . . . help them organize and defend themselves and fulfill agrarian legislation."[43] Little wonder they personified every-

41 Presidente de la Junta de Administración Civil to Secretario Gral. de Gobierno, December 23, 1922, AGEV, RF, Tierras, Deslindes.

42 Ibid.

43 Quotation from Hernández Fujigaki, "Los agrónomos frente a los retos nacionales," 188. Conversely, villagers saw *private* surveyors as a distinct threat to *their* interests. When private surveyor Jorge Macedo, with the agreement of the

thing Veracruz landlords feared. The fact that Tejeda, as well as another radical governor, Yucatan's Felipe Carrillo Puerto, trained and worked as land surveyors during the revolution simply confirmed their suspicions. Tejeda had been a surveyor in Chicontepec in 1913, when Huerta overthrew Madero; Carrillo Puerto attended the Escuela Nacional de Agricultura, the premier training ground for Mexican surveyors, and had worked on surveys in Morelos with the likes of Ignacio Díaz Soto y Gama (brother of Antonio, the head of the National Agrarista Party) and Fidel Velázquez, future head of the Regional Confederation of Mexican Workers.[44]

While not the sole ingredient, the surveyor was a critical one in every phase of the reform process. A pueblo deciding to petition for a grant of dotación would elect three citizens to serve as a committee to solicit the governor for a grant of land. The governor certified the community's "political validity" and then passed the solicitation on to the CLA, who in turn notified landowners of the solicitation. A surveyor then arrived to perform a *visita de inspección,* consisting of a detailed investigation into cadastral records, property archives, and tax offices, as well as interviews with local inhabitants, in order to arrive at a general idea of the nature of landholdings and land values in the area.[45] This work proved critical: such research often revealed landowners owned—or at a minimum, controlled—much more land than that listed in land records and titles.[46] Moreover, their investigations determined the size of landholdings in

state government, set out to survey and divide a number of large haciendas in central Veracruz into small properties for sale, various *comités agrarios* blocked his work, claiming that such lands should be incorporated into their ejido grants. Ingeniero Jorge Macedo to Gobernador del Estado, April 24, 1929, AGEV, RF, Tierras, Deslindes, exp. Lo relacionado con el fraccionamiento que está llevando a cabo en grandes latifundios de diversos lugares pertenecientes al Estado de Veracruz.

44 See Gómez, *Historia de la Comisión Nacional Agraria,* 186, 268; Womack, *Zapata and the Mexican Revolution,* 231–33.

45 Resumen de los asuntos que tiene que investigar el enviado especial que irá al Cantón de Papantla en vista de que durante nueve meses no se han podido obtener los datos necesarios para dictaminer la solicitud de ejidos, June 29, 1918, AGEV, CLA, caja 2, exp. 1918; and "Instrucciones," AGEV, CLA, caja 4. See also the summation in Craig, *The First Agraristas,* 249–58.

46 See, for example, the report of Gustavo Alvo Campos, October 29, 1932, AGEV, CAM, exp. 2101, Congregación de Romeros, Municipio de Las Minas, Dotación de ejidos.

order to prevent possible circumvention of the land reform by landowners who broke up their holdings into parcels small enough to avoid expropriation. Such strategic fragmentation was illegal, and landowners were warned that any attempts would be futile as all sales after a solicitation were regarded as null and void.[47] Still, landowners were not always unsuccessful. Due to the length of time it took to get surveyors into areas and to grant provisional rights, sales could go unnoticed for years, at which point the purchasers had been cultivating the land for nearly a decade. Ramón Beltrán, Lorenzo García, and Pablo Alvarez were all victimized by the Gómez family of Omealca in this way. The family sold them a number of plots, in the process fracturing their lands in order to avoid expropriation. Years later, when a surveyor finally arrived to mark the provisional grant, the three found themselves on the verge of expropriation and petitioned desperately to the governor for recourse.[48]

Having completed his investigations, the surveyor then surveyed the pueblo's urban area, existing communal lands, and both small and large private properties in the immediate vicinity, producing a final map that included notices on land classifications, roads, and irrigation works. He forwarded this map to the CLA, along with his field notes and a full report including a census, description of lands, average wages, types of cultivation, and commercial value of land by hectare for both the community in question and neighboring landholdings. The CLA then issued a ruling on whether or not to pursue the grant and, if positive, what the respective size of each parcel should be.[49] From there, the material passed to the governor for confirmation, at which point yet another surveyor

47 For example, see Presidente de la Comisión Nacional Agraria to Manuel Granda, June 20, 1923, AGEV, CAM, exp. 330, municipio de Las Minas, Poblado Zomelahuacan, Dotación de ejidos.

48 See Ramón Beltrán to Gobernador del Estado, May 7, 1932; Lorenzo García to Comisión Local Agraria, January 2, 1933; and Pablo Alvarez to Gobernador del Estado, August 3, 1934, all in exp. Omealca.

49 A "typical parcel" was a measurement denoting what size parcel would be needed for a head of family in the ejido. Fixing the size of a "typical parcel" was no easy feat because of the sheer number of variables involved. The Reglamento Agrario did in fact offer a series of guidelines, but the government quickly realized that such guidelines could hardly be realistically useful, and it notified surveyors that, while the agrarian census had to be rigorously determined, a "typical" parcel did not. See "Instrucciones," AGEV, CLA, caja 4, and Cuadros Caldas, *México-Soviet*, 257–61.

would be sent to give provisional possession to the community and eji-datarios could begin to cultivate. He would survey the perimeter of the ejido grant, create a "definitive map," and oversee the election of a local administrative committee. The case would then go to the CNA for final review and be forwarded to the president for the grant of final posses-sion. If affirmed, a last surveyor would go to the pueblo to confirm the survey and officially give final possession.[50]

Surveyors also wielded a significant amount of informal power, and it was this as much as anything else which landowners feared. Veracruz landowners perceived surveyors, especially those in the employ of the CLA, as favoring villagers at their expense. Given that Tejeda, as state governor, appointed CLA personnel, their impressions were not without foundation. That *agentes de propaganda* distributed small books with such titles as *Como podrán los pueblos obtener sus tierras* (The means by which pueblos can secure their lands) to villagers while surveyors kept them abreast of new legislation must also have angered many a land-owner.[51] Moreover, surveyors' reports often acquired the status of truth in regional and national offices, something that infuriated landowners. In 1925 a group of ex-*peones* (laborers) from the hacienda of San Diego in Acultzingo petitioned for a grant of land. Even though the Agrarian Law prohibited hacienda peons from petitioning for lands from the es-tates on which they worked, a surveyor in the employ of the CLA, Othón Aguirre, visited the region to examine their claim.[52] He created a census of some 401 inhabitants and ninety-two families and effectively granted them political status as a congregación. Dolores Rivadeneyra de Reyn-olds, the owner of the hacienda, immediately appealed to the CLA claim-ing that Aguirre had falsified the census. "The exact number of inhabi-

50 See Cuadros Caldas, *México-Soviet*, 257–61; and three articles in Comisión Nacional Agraria, *Recopilación Agraria:* "Instrucciones a que deben sujetarse los Ingenieros dependientes de la Comisión Nacional Agraria y Comisiones Locales Agrarias," 84–92; "Circular de 10 de octubre de 1922, sobre los procedimientos topográficos que deben seguirse para expeditar las posesiones provisionales," 93–95; and "Instrucciones para los Ingenieros que dependen del Departamento de Aprovechamiento de Ejidos de la Comisión Nacional Agraria," 216–18.

51 See Daniel Muñoz Esteban to Presidente de la Comisión Local Agraria, May 21, 1918, AGEV, CLA, caja 4, exp. de Misantla. See also Gudiño and Palacios, "Peticiones de tierras," 104.

52 On the Agrarian Law restriction, see Nugent, *Spent Cartridges of Revolu-tion*, 90.

tants," she claimed (with little precision), "was 50 or 60." Moreover, she argued their combined living quarters hardly constituted a "congregación, Rancho, or Ranchería" and thus they could not even petition for an ejido in the first place; finally, she demanded that Aguirre be indicted for falsifying the information given to the CNA.[53] Her demands were dismissed. Whether Aguirre actually falsified his information is less relevant than the implication of power to which the accusation alludes. Reynolds clearly saw him as the root of her problems.

As did some powerful individuals *within* communities. At times, petitioners had to fear not only landlords irate at the immanent inversion of their world but also their own municipal authorities. Municipal officials, many of whom were themselves powerful landowners, feared the impact an ejido might have on their own power and property. Manuel Jiménez Rodríguez held the position of síndico in Tonayán in 1921. He also happened to be one of the landowners accused of taking village land as well as the legal representative of Faustino Vázquez, a former village síndico and now the largest landowner in the region.[54] Municipal authorities elsewhere could be found impeding surveyors' work, attacking villagers who had petitioned for an ejido, or siding with large landowners in land conflicts.[55] Other village members used the land reform as a means to sustain or expand their own power. Hacendados may have been the most obvious of antagonists, but as postrevolutionary recon-

53 Dolores R. Reynold to H. Comisión Local Agraria del Estado de Veracruz, October 14, 1926, AGEV, CAM, exp. 364, Municipio Acultzingo, Población Potrero de San Diego, Dotación de Ejidos. The categorization of settlements was extraordinarily subjective and could often work in favor of hacendados. See Cambrezy and Marchal, *Crónicas de un territorio fraccionado*, chap. 1.

54 See Vecinos del Municipio de Tonayán to Gobernador del Estado, December 31, 1917, Tonayán, exp. 72; El Presidente de la CLA to Agente Manuel Jiménez Rodríguez, May 23, 1918, Tonayán, exp. 72; El Presidente de la CLA to Presidente Municipal de Chapúltepec, August 31, 1918, Tonayán, exp. 72; and Salvador de Gortari to Ing. Othon Aguirre, October 10, 1921, Tonayán, exp. 72.

55 The documentary record is full of such instances. For illuminating examples, see Presidente del Comité Ejecutivo Agrario, Epistacio Arao[?], to Gobernador del Estado, March 18, 1925, AGEV, RF, Tierras, Deslindes, exp. Lo relativo a la queja que presenta en contra del señor Antonio Romero; Presidente de la Liga de Comunidades Agrarias to Presidente de la Comisión Local Agraria, April 15, 1926, AGEV, CLA, caja 4, exp. Límites relativo al Municipio de Acatlán; and Mateo Rodríguez to Procurador de Pueblos en el Estado, July 21, 1928, exp. Omealca.

struction unfolded, village elites also jockeyed for power. Reform did not necessarily level social hierarchies in communities. In Acultzingo a small number of villagers—Félix Martínez, Crecencio Andrade, and members of the Alvarado family—had managed, during the revolution, to wrest control over significant chunks of village land from the owners of the neighboring rancho of San Isidro and the hacienda of San Diego. Their acquisitions included about one hundred hectares from the latter, for which they later became known by poorer villagers as the *socios de las cien ectarias* (the one hundred hectare partners, or buddies).[56] Already by 1920 these individuals were large landowners in their own right and they signed the petition for restitution in 1921 in order to ensure their dominance over village land and politics.[57] Their machinations eventually raised the ire of others in the community when they attempted to acquire exclusive control over certain waterways in the village, with the support of the municipal president. This prompted a letter from some twenty-five villagers to the governor asking that he not grant any individual rights to the waterways within the municipal boundaries. Let any document, they wrote, favor "everyone in the pueblo and not just a few individuals."[58] Behind a communal facade, village elites and authorities could use the reform as a means to consolidate their own power, mask their dominance, and expand their holdings.

Faced with such contextual intricacies, agrarian bureaucrats looked to the surveyor for a solution. "Surveyors," the president of Veracruz's CLA wrote in 1925, "are in the most intimate contact with the campesino

56 Varios vecinos de Acultzingo to Gobernador del Estado de Veracruz, Sr. Adalverto Tejeda, May 27, 1929, AGEV, RF, Sección de Aguas caja 66, manantiales, Acultzingo; and Plano de la Fracción de terreno rematada a la Hacienda de San Diego, ubicada en Acultzingo, Ver., a favor del Sr. Cresencio Andrade, November 23, 1929, AGEV, Planoteca, Acultzingo. The reference to "cien ectarias" is a sardonic one, referring to Article 249 of the Agrarian Law of January 6, 1915, which established that irrigated land that did not exceed one hundred hectares was not subject to expropriation. Powerful, propertied individuals in Acultzingo thus managed to stay just under the limit that would have subjected them to expropriation.

57 Pueblo del Acultzingo to Gobernador del Estado, March 18, 1921, AGEV, CAM, exp. 247, Municipio Acultzingo, Poblado Acultzingo, Dotación de ejidos [hereafter exp. Acultzingo].

58 Varios vecinos de Acultzingo to Gobernador del Estado de Veracruz, Sr. Adalverto Tejeda, May 27, 1929, AGEV, RF, Sección de Aguas, caja 66, manantiales, Acultzingo.

because of the nature of their work" and are thus best able to prevent campesinos from being victims of the "exploitations to which they are subject by their so-called leaders."[59] Perceiving the surveyor as the best means by which to circumvent resistant municipal officials and traditional structures of local power, instructions and correspondence for peasants related to an ejido grant were often forwarded through the surveyor rather than the municipal president.

Little wonder, then, that revolutionary muralists like Diego Rivera valorized surveyors (figure 19).[60] To the new promoters of postrevolutionary culture, Cándido Aguilar may have simply been looking in the wrong place for "honorable revolutionaries." Whether or not the appellation is appropriate, one thing is certain: surveyors were the medium through which many rural people *experienced* the new revolutionary state.[61] Like the revolutionary schoolteacher, the surveyor appeared as the most concrete manifestation of a revolution still in the making and became the object of competing interventions.[62]

III. DRIP IRRIGATION

Opponents of the reform had more than Carrancista generals on their side: they had time. Reform progressed in lethargic fashion, even after Alvaro Obregón replaced Carranza in the presidential palace. The lack of qualified surveyors that slowed the pace of the land divisions in Porfirian Mexico also plagued the process of land reform after the revolution. With an appropriate agrarian metaphor, one CNA member at the time described the process as "drip irrigation" (*riego por goteo*).[63]

59 José Gutiérrez to Ing. Jorge Vizcaino, March 20, 1925, exp. Omealca.

60 The image presented here is only one of an array of images by Rivera celebrating surveyors, agronomists, and other agrarian bureaucrats. See, for example, the various images he provided for the bound volumes of proceedings of the Liga de Comunidades Agrarias y Sindicatos Campesinos de Tamaulipas. See Gómez, ed., *Primera convención*; Gómez, ed., *Segunda convención*; and Gómez, ed., *Tercera convención*.

61 Akhil Gupta has observed that it is precisely at the lower levels of the bureaucratic hierarchy that the majority of people experience "the state." Gupta, "Blurred Boundaries."

62 On rural schoolteachers, see Vaughan, *Cultural Politics in Revolution*, and Rockwell, "Schools of the Revolution."

63 Gómez, *Historia de la Comisión Nacional Agraria*, 239.

FIGURE 19. Honorable revolutionaries: Diego Rivera, untitled sketch from *Primera Convención de la Liga de Comunidades Agrarias y Sindicatos Campesinos del Estado de Tamaulipas*, 1926.

Even that metaphor appeared optimistic. The genius of drip irrigation is that it supplies just enough moisture to meet the requirements of a given crop. Yet the CLA could hardly manage to supply the surveyors necessary to meet all of the demand for land surveys and grants. A cursory perusal of petitions to the CLA for restitution or dotación in postrevolutionary Veracruz reveals a single phrase appearing with near mantric regularity: "as soon as there is a surveyor available." In the days leading up to and following Carranza's promulgation of the new Agrarian Law, communities aggressively pursued land restitutions and grants and by 1919 the CLA was already overwhelmed.[64] That year, the interim president of the CLA reminded the governor that "the rapid resolution of all the affairs with which this office is in charge is directly related to the number of surveyors at its disposal. To decide upon a land case, it is imperative that a specially commissioned surveyor collect data and information and create maps or sketches of the lands that have to be returned or granted to communities."[65] He suggested that the three propaganda agents currently in the employ of the CLA be replaced by two surveying assistants. "The agents," he claimed, "have done nothing of importance," whereas additional surveyors could speed up the work dramatically. His antipathy toward the agents may have resulted from the fact that at least one of them, Daniel Muñoz Estéban, was a large landowner and ally of General Sánchez. Indeed, when de la Huerta's rebellion came in 1923, Muñoz would organize peasant irregulars against Tejeda.[66]

Work progressed slowly also because surveyors lacked the full complement of materials necessary to do their surveys and thus shared, or appropriated, each other's equipment. The case of Pedro Valero is illustrative of the dismal situation. He delayed his field assignments because a colleague had borrowed one of his measuring chains and theodolite;

64 The villages of Ursulo Galván, Acatlán, Coacoatzintla, and Chapultepec all petitioned in January 1915 for either restitution or a grant of land. For Chapultepec, see AGEV, CLA, caja 2, exp. Chapultepec, Límites, 1915; for the others, see *Secretaría de la Reforma Agraria: Perfil agrario del Estado de Veracruz: Delegación Xalapa, 1915–1979.* This catalog was never published. Ingeniero Héctor Rivadeneyra, a retired surveyor who worked in the Veracruz SRA, generously loaned me his personal copy for consultation.

65 El presidente interino de la Comisión Local Agraria to Gobernador del Estado, October 11, 1919, AGEV, CLA, caja 3, exp. Personal en conjunto de esta Comisión Local Agraria.

66 Fowler-Salamini, *Agrarian Radicalism in Veracruz,* 43–44.

his other chain had been broken on a previous survey; and another colleague had borrowed both his protractor and his geographic coordinate sheets.[67] Valero's situation may have provided the inspiration for a state decree, issued that same year, requiring that "all the engineering instruments and drawing implements which the Department [CLA] provides to the mentioned surveyors for their work, stay under the sole and exclusive responsibility of them . . . [The surveyors will be] financially and personally responsible for the voluntary or involuntary loss or deterioration of them."[68]

Such problems, following on years of Carrancista ambivalence, exasperated petitioners. Increasingly, they threatened to take matters into their own hands. Petitioners resorted to thinly veiled threats of new rounds of violence and disorder, a sure means by which to jolt the administration into action. The municipal president of Tonayán in early 1923 notified state authorities that land issues between villages in the sierra needed to be addressed soon "in order to maintain Peace in our land because if the difficulties which all the Pueblos have over their lands are not repaired, the case may or could arise in which they will move against each other and we do not want this to happen." A quick resolution, he concluded, would ensure that "tranquility will reign in our region."[69] That same year, the owner of the hacienda of San Diego attempted to forestall the return of five ojos de agua to the municipal control of Acultzingo, suggesting to a judge in Orizaba that the situation might require the use of federal forces.[70] The municipal president responded aggres-

67 Inventario de instrumentos y útiles de ingeniera de la Comisión Local Agraria del Estado, que quedan desde el día de la fecha a cargo del Dibujante de la expresada Oficina, Ingeniero Pedro L. Valero, February 21, 1919, AGEV, CLA, caja 3, exp. Muebles, instrumentos y útiles relativo a la Comisión Local Agraria, 1919.

68 El Sec. de Gobierno to Presidente de la Comisión Local Agraria, March 1, 1919, AGEV, CLA, caja 3, exp. Muebles, instrumentos y útiles relativo a la Comisión Local Agraria, 1919.

69 Presidente Municipal Luis Z. Fernández to Secretario General de Gobierno, January 8, 1923, AGEV, RF, Tierras, Deslindes, exp. Tonayán: Referente al DESLINDE de terrenos entre ese Municipio y sus limítrofes. For a similar threat, see Presidente de la Junta de Administración Civil de Acultzingo to Gobernador del Estado, October 21, 1930, AGEV, RF, Tierras, Límites, exp. Soledad Atzompa.

70 Dolores Rivadeneyra de Reynold to C. Juez 10. de 1a. Instancia de Orizaba, June 5, 1923, AGEV, RF, Sección de Aguas, caja 52, exp. Amparo que promovió ante el C. Juez Primero de Primera Instancia.

sively: "If a federal force comes to execute the order of the Judge, there could be a conflict like that at Puente Nacional, because it is also known that there is a volunteers corps [*cuerpos de voluntarios*] in this municipio, all Indians, and the entire pueblo will rise up and I will be unable to contain them without risking my life." "Besides," he concluded with more cheer, "for now no one is lacking for water because there has been plenty of rain."[71] Acultzingo assumed control over the ojos that very year.

Still, it took some time for their *land* petition to progress. By 1933, after more than a decade of patience, they threatened to do the survey themselves, a statement that compelled the CLA to remind them that "only a certified engineer can undertake the survey of the ejido lands."[72] Others by this time had employed similar tactics. In 1929 agraristas in Cosamaloapan assured the governor that "the entire pueblo has complete confidence that after a struggle with Capital the orders of the Revolution will be carried out," but such lip service could hardly veil their disillusion, and they swore that if a surveyor did not arrive soon "we will be obliged to do the survey ourselves."[73]

Such disenchantment threatened the provisional peace and tentative legitimacy of early postrevolutionary Mexico and forced federal officials to respond accordingly. Tejeda personally intervened to try and speed up the processing of petitions to the CLA. Although evidence is scant, the backlog may have also convinced the government to reemploy ex-CGE surveyors, like Francisco Canóvas, who only years earlier had been

71 Guadalupe Martínez to Gobernador del Estado, June 21, 1923, AGEV, RF, Sección de Aguas, caja 52, exp. Amparo que promovió ante el C. Juez Primero de Primera Instancia. The reference to Puente Nacional was to an infamous incident in Veracruz, in the spring of 1923, in which guardias blancas attacked and kidnapped the president of an agrarian committee upon his arrival at the municipal hall in the town of Puente Nacional. Tejeda protested to both the Ministry of War and the CNA with little result and thus ordered the Civil Guard to bring the perpetrators to Xalapa to face trial. Upon arriving in Xalapa a shoot-out ensued in which seven people were killed. President Obregón sided with General Guadalupe Sánchez against Tejeda and ordered the Civil Guard disbanded. See Fowler-Salamini, *Agrarian Radicalism in Veracruz*, 38–39.

72 El Comité Particular Administrativo Agrario de Acultzingo to Presidente de la Comisión Local Agraria, August 29, 1933 exp. Acultzingo; Pablo Hernández to Andrés Montero y Silverio Díaz, September 21, 1933, exp. Acultzingo.

73 Letter included in Gobernador del Estado to Ingeniero Delgado de la Comisión Nacional Agaria, October 5, 1929, AGEV, CLA, caja 5.

FIGURE 20. Exile's return: Francisco Canóvas y Pasquel (second from the left) and family dining in their Xalapa home with governor of Veracruz, Heriberto Jara (center). Date unknown but ca. 1924–1927. Courtesy of Carmen Boone de Aguilar and Daniela Canovas Rebling.

demonized as dangerous reactionaries to the revolutionary cause (figure 20). Despite this, Tejeda's interventions and personnel increases had minimal effect. The huge backlog and growing disillusionment compelled CLA president de Gortari to order surveyors to "carry out rapidly the survey and marking of the ejido . . . with topographic measurements of little precision," just adequate to prevent disputes with affected property holders.[74] Such a strategy had proven problematic in the past. In 1919 the CNA rejected a provisional grant to a village because of imprecision in the survey.[75] In other instances, surveyors flatly stated that they could not submit the necessary materials that quickly if they were to

74 de Gortari to Aurelio Ortega, December 14, 1931, Tonayán, exp. 72; see also de Gortari to Ing. Jesús Medina, May 19, 1931, exp. Omealca.

75 Presidente interino to Gobernador del Estado, December 9, 1919, AGEV, CLA, caja 3.

avoid "irreparable errors."[76] By 1930 there were some 1,109 new petitions for land waiting to be processed and some 645 existing petitions
still pending, which de Gortari estimated would alone take some three
years to resolve.[77]

In sum, if the land reform was "drip irrigation," many communities
were left out to dry. Delays in the land survey wrought dramatic effects
on the villagers of Omealca, who, with no provisional permission to the
land, remained at the mercy of the Gómez family.[78] "The terratenientes
[large landowners] have not stopped pressuring us," one villager wrote,
and people were being forced to migrate away to find work. "Send a surveyor to do the work," he pleaded, "and to give us provisional possession
of our lands so that we can subsist with our work and educate our children."[79] Between 1928 and 1930 no fewer than seven surveyors were
ordered to go to Omealca to carry out a survey, but few either went or
stayed.[80] The combination of powerfully entrenched local elites, armed
and angry, and a queue of cases meant surveyors could avoid the most
troublesome spots in the countryside.

IV. THE TIME AND SPACE OF *REIVINDICACIÓN*

The pace of reform, while clearly affected by personnel questions, also
progressed slowly because campesinos, landowners, and state officials
struggled over the very meaning of revolution and reform. If revolutionary surveyors and agrarian bureaucrats viewed reconstruction and the
agrarian reform as the dawning of a new era, for many villagers it was a
long awaited reivindicación—a time of both restoration and vindica-

76 Ortega to de Gortari, October 15, 1932, Tonayán, exp. 72.

77 Fowler-Salamini, *Agrarian Radicalism in Veracruz*, 99.

78 Pascual Cervantes, Presidente del Comité Particular Ejecutivo, to José Gutiérrez, Presidente de la Comisión Local Agraria, February 7, 1925, exp. Omealca;
Vicente Hernández, Celso Martínez, and Juan Morales to Comisión Local Agraria, August 23, 1929, exp. Omealca.

79 Vicente Hernández, Presidente del Comité Agraria de Omealca, and
Agustín Macías, Presidente del Comité Agraria de Xuchiles, to Salvador de Gortari, included in de Gortari to Ing. Othón Aguirre, October 4, 1929, AGEV, CLA,
caja 5.

80 See the collected materials in Omealca. Tonayán had a similar experience;
see the documents in Tonayán, exp. 72.

tion. For villagers from Tonayán in 1917, the agrarian law meant nothing less than the "blessed restoration of the rights of the oppressed."[81] A few years later, the municipal president of Santiago Huatusco, a town in the cane-growing canton of Córdoba, was even more direct: "The voice of the proletariat is no longer drowned out by the sugary words of the magnates. . . . The time of reivindicaciones has arrived."[82]

Reivindicación meant restitution. Restitution, one of the two forms by which campesinos could petition for land under the agrarian reform law, signified the return of lands taken illegally in the past, based upon an extensive examination of a village's historical documentation. The alternative manner of acquisition, known as dotación, was to petition the state for a grant of land in the form of an entirely new spatial construct created out of holdings of large landowners in the immediate vicinity. That many petitioners opted to petition for restitution is not remarkable. The revolution was fought, as Daniel Nugent and Ana María Alonso have argued, not only over control of the means of production but the production of meaning.[83] Land was more than space: it was history, rights, and tradition—it was place. Reivindicación and restitution signified the return of lands but also validated a community's historical and "sacred rights" as a pueblo and *municipio libre* (free or autonomous municipality).[84] Reivindicación could not be captured in the calculus of acreage, devoid of cultural meaning; it was to be measured in history, not hectares.

Thus, for petitioners from Acultzingo, restitution of lands and waterways meant "justice" for a "pueblo founded prior to the Colonial Gov-

81 Vecinos de Municipio de Tonayán to Gobernador del Estado, December 31, 1917, Tonayán, exp. 72.

82 Presidente Municipal de Huatusco to Gobernador del Estado, February 20, 1922, AGEV, RF, Tierras, Deslindes, exp. Lo relacionado con los límites entre los Municipios arriba indicados [hereafter exp. Huatusco].

83 Nugent and Alonso, "Multiple Selective Traditions," 246. Daniel Nugent's observations culled from his work in Namiquipa, Chihuahua, are suggestive: "Mexicans who petitioned for land often wanted *their own* land. . . . Securing a restitution was justice, while securing a *dotación* was an insult, a matter of the state posturing as a *patrón*." Nugent, *Spent Cartridges of Revolution*, 91. Emphasis in the original.

84 Isidro Acosta, Jesús Sánchez et al. to Gobernador Consitucional del Estado, March 7, 1930, exp. Huatusco.

ernment and recognized and granted status by the viceroy don Luis de Velasco" in 1559.[85] Villagers from Tonayán wanted lands they claimed rights to as a pueblo "since time immemorial" (although they specified a founding date of 1540), offering proof in the form of titles and "an ancient map with hieroglyphic characters, which express clearly that we had great extensions of land as Ejidos."[86] The municipal authorities of Santiago Huatusco, in their request for restitution, attached their rights to a national history: "The Governor of our Nation, which then had the name of New Spain, the señor Viceroy Don Luis de Velazco, in compliance with the Reales Ordenanzas, gave this pueblo as a Republic or community of Indians (which in the pre-Cortesian period was the ancient cacicazgo of Cuahtocho, destroyed, according to history, by the Mexican kings Axayacatl and Moctecutzoma Ilhuicamina) an extension of territory with the following borders."[87] After delineating the boundaries through references to landmarks and directions, the narrative tracked a variety of expropriations by hacendados over the course of the centuries, ending with a final land grab in 1885 by Porfirio Díaz's minister of war, Pedro Hinojosa. "The protests of the victims were for nothing," the letter concluded. "[T]he Judges acted deaf and the dispossession was consummated."[88]

Large landowners were not the only antagonists. Encroachment by neighboring villages often inspired petitions for restitution. Petitioners from Acultzingo sought not only the return of lands taken by neighboring hacendados but also an end to conflicts over liminal land with the neighboring village of Cañada de Morelos. Villagers from Tonayán claimed they were victimized not only by their own Faustino Vázquez but by the villages of Tlacolulan and Chapultepec.[89] Indeed, for one official there, the reform law was clear enough: after four years of waiting, he told the governor to "comply with the laws offered us in the programs

85 Pueblo de Acultzingo to Gobernador del Estado, March 18, 1921, exp. Acultzingo.

86 Vecinos del Municipio de Tonayán to Gobernador del Estado, December 31, 1917, Tonayán, exp. 72.

87 Presidente Municipal de Huatusco to Gobernador del Estado, February 20, 1922, exp. Huatusco.

88 Ibid.

89 Vecinos del Municipio de Tonayán to Gobernador del Estado, December 31, 1917, Tonayán, exp. 72.

of the Constitutionalist revolution; return to the Pueblos the lands invaded by others."[90] Municipal authorities *accused* of poaching neighboring lands countered with their own understanding of reform, one that tended to adopt the language of dotación rather than restitution. For example, authorities in Naolinco, confronted with charges of having illegally acquired a large section of a neighboring village's land, argued that "the spirit of the Agrarian Law 'is to give lands to the rural populations that lack them,' not to expropriate the lands of those that already have them."[91]

Restitution, in sum, rested upon a language of usurpation and tyranny, of time immemorial and historical solidarity. The vocabulary of restitution conjured the community as a historical entity intricately tied to the history of the nation rather than indebted to the revolutionary state. "It is on the long corruption," wrote Raymond Williams, "not the lucky exception that the landless insist."[92] Although writing of the historical vision of England's rural poor in the era of enclosures, Williams's eloquent words resonate profoundly with the vision of Mexico's rural cultivators contemplating postrevolutionary reform. Premised upon their historical rights, rooted in a nationalist discourse, pueblos insisted on the long corruption that had plagued their community and the nation, rather than allowing themselves to be portrayed as the submissive recipients of state grace in the form of dotación.

Dotación, after all, effectively reduced the revolution to a matter of material deprivation. This is not to suggest that material concerns were unimportant. Francisco Canóvas, returned from exile and working as a surveyor for the federal government, painted a terribly moving portrait—one worth quoting in full—of the reality in which much of Veracruz's serrano population lived when he visited the municipality of Soledad Atzompa. They lived in "appalling misery and poverty" because of

the lack of cultivable land, as nearly all of the lands which they currently possess are very mountainous and those less so are of clay soil,

90 Presidente de la Junta de Administración Civil to Jefe del Depto. de Fomento y Agricultura, February 8, 1922, AGEV, RF, Tierras, Deslindes, exp. Tonayán: Referente al deslinde de terrenos entre ese Municipio y sus limítrofes.

91 Letter dated November 27, 1920, AGEV, RF, Tierras, Deslinde, exp. San José Miahuatlan, Ex-Cantón de Xalapa.

92 Williams, *The Country and the City*, 42.

not amenable to cultivation[;] and a drought, even a short one, or a year of abundant rainfall destroys their crops. This lack of land appropriate for cultivation requires them to look elsewhere for their means of existence, and they thus make cookware, chairs, little tables, window boxes, and other wood products, all very common, which they carry on their backs [and] go to surrounding towns to sell at absurdly low prices. With what they make they buy corn and beans. They also make coal that they carry like mules for many leagues to make a few centavos.[93]

He summed up his report by emphatically arguing that the inhabitants should receive a dotación. But the series of letters from the municipal president, Vicente Francisco, which had prompted his visit, made no reference to a land grant. Rather, Francisco sought *restitution* of lands seized by neighboring villagers prior to and during the revolution. "*Our land* has remained in the power of two groups of outsiders and our suffering continues," Francisco told the state governor.[94]

Canóvas was not exceptional in his conflation of restitution with dotación. In the revisionist lens of the postrevolutionary vanguard, an expressed desire for reivindicación could quickly be mistranslated into a need for material improvement. In revolutionary manifestos, in the social science literature dedicated to the "agrarian question," and in the words of political ideologues, villagers frequently appeared in the passive, as "motives" for change and "objects" of liberation, indebted to the new revolutionary state rather than the foundation of its legitimacy.[95] The process of reform, often hashed out in metropolitan offices by individuals with little immediate link to popular movements, was beleaguered by a similar set of presumptions.[96] References to hectares and soil classifications punctuated the conversations among members of the CLA and CNA. While clearly the size of grants and quality of land was of crucial importance to petitioners, such an emphasis reduced pueblos to a quantitative episteme of use largely to a bureaucracy attempting to meet

93 Francisco Canóvas to Jefe del Depto. de Comunicaciones y Obras Públicas, Jalapa, July 2, 1932, AGEV, RF, Tierras, Límites, exp. Soledad Atzompa.

94 Vicente Francisco to Gobernador del Estado, November 25, 1930, AGEV, RF, Tierras, Límites, exp. Soledad Atzompa. My emphasis.

95 See Gudiño and Palacios, "Peticiones de tierras."

96 Nugent and Alonso, "Multiple Selective Traditions," 228–29.

the "economic needs" of petitioners while treating the "affected land-owners . . . with justice."[97]

A discussion between two CLA officials in Veracruz in 1918 provides a case in point. Responding to a request for restitution of lands from the village of Chiltoyac, Salvador de Gortari and Guillermo Rebolledo dis-cussed how much land should be returned. Rebolledo suggested a dota-ción (*not* a restitution) of 1,350 hectares. De Gortari in turn argued that the dotación should at least be equal to the amount requested in the original petition for restitution, some 1,755 hectares, because people who had left the village in the past to find work might return if they heard lands were being redistributed. They reached a compromise of 1,500 hectares but the shift to dotación was never questioned.[98]

In fact, both federal and state governments systematically favored land grants over demands for restitution. Nationally, between 1916 and 1940 only 6 percent of land distributed through the agrarian reform was granted through acts of restitution. If one were to extend this through to 1980, the figures are even more dramatic: a mere 1 percent of all land distributed nationally was restituted land.[99] Why was this so, par-ticularly given that many petitioners in the 1910s and 1920s explicitly sought restitution?

Restitution constituted enormous problems for the agrarian bureau-cracy. For one thing, it often proved to be incompatible with dotación. When municipal authorities petitioned for restitution, they frequently requested the return of *all* lands within what they perceived to be their historical jurisdiction since time immemorial. The petitions for restitu-tion from Tonayán, Acultzingo, and Santiago Huatusco all came from the municipal seats (cabeceras). But what then would become of the var-

97 Consideraciones generales sobre el objeto de los trabajos topográficos de la Comisión Nacional Agraria, AGEV, CLA, caja 4. See also AGEV, CLA, caja 3, exp. Co-pias de actas de la Comisión Local Agraria, 1919, Acta 116, October 15, 1918; and Nugent and Alonso, "Multiple Selective Traditions."

98 AGEV, CLA, caja 3, exp. Copias de actas de la Comisión Local Agraria, Acta 116, October 15, 1918.

99 For the two figures see, respectively, Nugent, *Spent Cartridges of Revolution*, 181n17; and Sergio Zendejas, "Appropriating Governmental Reforms." Veracruz boasts similar percentages. See *Secretaría de la Reforma Agraria: Perfil*. The fig-ures are contestable in their specificity—for example, one author has suggested that through 1980 17 percent of all resolutions were restitutions—but the broader point is sound.

ious settlements and subject towns within the municipal borders?[100] How could land be restituted to the cabecera of a municipality and at the same time be granted through dotación to a *sujeto* (hamlet or village) within its confines? These were precisely the kinds of complications CLA authorities tried, at first, to negotiate. The CLA president in 1917 received a request for restitution from the cabecera of Chiconquiaco. In response, he asked if they wanted "restitution of land to work in common or the return of land to your political jurisdiction."[101] The difference was critical: the former could impact grants of dotación whereas, in theory, the latter would not. But such a distinction did not hold in practice and did little to resolve the paradoxical nature of restitution and dotación.

Restitution and dotación were also complicated by the politics of agrarismo (agrarian radicalism closely associated with Tejeda's administration). A conflict between the municipalities of Santiago Huatusco and San Juan de la Punta is a case in point. The authorities of San Juan petitioned for a dotación in 1921.[102] Local lands available for expropriation were those of the hacienda Trapiche, most of which were located within the municipal limits of Santiago Huatusco. Out of fear that they would lose control over that land, Santiago Huatusco petitioned for restitution in 1922. Disputes immediately erupted over the location of municipal boundaries, and the CLA ordered Carlos Olivares—an ex-CGE employee now working for the CLA—to the area to gather information, even though CLA officials had emphatically stated that questions regarding municipal limits were outside their purview.[103] Olivares had little

100 See Aboites and Morales Cosme, "Amecameca, 1922," and Gómez, *Historia de la Comisión Nacional Agraria*, 96.

101 Presidente de la Comisión Local Agraria to Presidente del Ayuntamiento de Chiconquiaco, October 8, 1917, AGEV, CLA, caja 2, exp. Documentos del Municipio de Chiconquiaco, Cantón de Jalapa.

102 See the history recounted in Salvador de Gortari to Secretaría General del Gobierno, May 20, 1930, exp. Huatusco.

103 Presidente Municipal de Huatusco to Gobernador del Estado, August 22, 1922, exp. Huatusco; Presidente Municipal de Huatusco to Gobernador el Estado, September 13, 1922, exp. Huatusco; Presidente Municipal de San Juan de la Punta, Pedro Olguín, to Subsecretario del Gobierno del Estado, November 8, 1922, exp. Huatusco. For instances where officials claimed that municipal limits were not part of their job, see Presidente de la Comisión Local Agraria to Presidente Municipal de Tonayán, January 30, 1923, Tonayán, exp. 72; Presidente de la Comisión

success, even after interviews with four men entrusted with preserving the knowledge of the village's 1824 boundary delineation (each of whom could recount no fewer than twenty former or existing mojoneras or points of orientation by name and in order as if one were walking them) and a search for the archive of his former employer, the Comisión Geográfico-Exploradora.[104] Over the course of the next decade, armed encounters and conflicts erupted and dissipated as CLA officials attempted to resolve the case. In the end, regardless of claims by authorities in Santiago Huatusco that the government would be committing "a grave infraction of the Constitutional Laws that govern us" if they did not respect their territorial claims, San Juan received a provisional grant of some seven hundred hectares. The vast majority of this land was located within the municipal boundaries of Santiago Huatusco.[105]

For the authorities of Santiago Huatusco, the outcome was predictable. Authorities in San Juan were apparently closely allied with Governor Tejeda and were members of the League of Agrarian Communities and Peasant Syndicates of the State of Veracruz (i.e., *agraristas*).[106] Indeed, inhabitants of Santiago Huatusco claimed that villagers from San Juan had organized "under the flag of agrarismo" for no other reason than to take their lands.[107] They then extended their argument through to its logical, and politically risky, conclusion: that they had experienced more justice during the Porfiriato than now as at least under don Porfirio they had been able to defend their holdings based upon their titles

Local Agraria to Presidente del Ayuntamiento de Chiconquiaco, October 8, 1917, AGEV, CLA, caja 2, exp. Documentos del Municipio de Chiconquiaco, Cantón de Jalapa.

104 Resumen de los testimonios, November 6, 1922, exp. Huatusco; Carlos Olivares to Subsecretario del Gobierno, December 6, 1922, exp. Huatusco; Gobernador del Estado [de Veracruz] to Secretaría de Agricultura y Fomento, January 11, 1923, AGEV, RF, Sección de Geografía y Estadística, caja 202, año 1923, Asunto: Suplicándole que devuelve a este Gobierno el ARCHIVO que fue facilitado en calidad de préstamo a la Comisión Geográfico-Exploradora que estuvo instalada en la Capital del Estado.

105 Presidente Municipal de Huatusco to Secretaría General del Gobierno, December 22, 1925, exp. Huatusco; Salvador de Gortari to Secretaría General de Gobierno, May 20, 1930, exp. Huatusco.

106 See Isidro Acosta, Jesús Sánchez et al. to Gobernador Constitucional del Estado, March 7, 1930, exp. Huatusco.

107 Ibid.

and maps. Municipal authorities were more careful: they wanted to avoid appearing antagonistic toward agrarismo. Juan Gallardo, a municipal council member in Santiago Huatusco, emphasized: "I am not an enemy of Agrarismo. I have been one of those who have praised entirely the reforms applied to our laws in favor of the proletariat . . . I do not speak as an enemy of Agrarismo, nor as an enemy of the Agrarian Committee of San Juan de la Punta. I speak as an *hijo of this pueblo*."[108]

For Gallardo, it seemed clear enough that villagers should not be permitted to petition for a grant of ejido land from holdings located in another municipio. After all, he mused, once having been granted lands within their municipal bounds what was to prevent authorities from San Juan having those very lands actually annexed to their municipality? Indeed, he suggested there was already precedent for such reconfigurations.[109] Although CLA president de Gortari stated that a dotación would not impact municipal jurisdiction, Gallardo's fears were substantiated when de Gortari pushed to have the legislature rectify the municipal boundaries so that San Juan's newly granted ejido lands would rest within their municipal boundaries.[110] Santiago Huatusco's fixation with restitution and its 1824 municipal borders conflicted with the new postrevolutionary state's desires to distribute land to the landless and the loyal.

Even under less politically volatile circumstances, restitution proved problematic. The process was a lengthy one, requiring the submission of historical documentation to the CLA and a review of the material by a paleographer attesting to its authenticity. Communities requested that lengthy searches be done at the Archivo General de la Nación for old titles and maps; the paleographers' reviews of existing material could take months, as could officials' attempts to wade through colonial documents. In addition, the very questions that had plagued Porfirian projects and officials with exasperating frequency now haunted the new agrarian bureaucracy: a lack of adequate, reliable documentation corroborating claims to prior possession; confusions and contradictions in the existing documentation; and a disjuncture between what appeared

108 Moción de Regidor Juan Gallardo, Ayuntamiento del Municipio de Santiago Huatusco, January 25, 1929, exp. Huatusco. Emphasis in the original.

109 Ibid.

110 Salvador de Gortari to Jefe de la Zona Ejidal, Luis Carrasco, June 3, 1930, exp. Huatusco.

in documents and what appeared on the ground. Even when materials were available and passed muster with paleographers, the process then required that a surveyor find and map the boundary markers and lines referred to in the titles and maps.[111] Such a survey could be both technically and socially arduous, as Olivares's experiences in Santiago Huatusco attested. Moreover, confronted with a growing number of cases and a limited number of surveyors, CLA officials could hardly endorse the kinds of instructions Emiliano Zapata reportedly gave the surveyors in Anenecuilco: "You engineers sometimes get stuck on straight lines, but the boundary is going to be the stone fence [which was the community's original boundary], even if you have to work six months measuring all its ins and outs."[112] For Zapata and his followers, restitution was not written with straight lines. These were incapable of winding themselves around "customs and usage," following the contours of history or conforming to the local impression of justice. Straight lines were state lines.

But it was one thing to carefully measure the lands of Zapata's followers in Morelos in 1915; quite another to do the same for all villages in Veracruz in 1925. For an agrarian bureaucracy suffocating under the crush of claims, straight lines were quick routes to completing surveys and accelerating the reform process, something de Gortari clearly understood: in 1931, confronted with a huge backlog, he ordered surveyors to map out land grants with the minimum amount of exactitude necessary to prevent disputes.[113] How much easier and quicker it would be to lay out an entirely new spatial unit of geometric simplicity through dotación than to try to follow the contours of memory and the faded curves of colonial maps in a process of restitution.

In sum, restitution severely complicated state officials' capacity to control the pace, manner, and parameters of reform. Clearly there was no single reason as to why they systematically favored land grants in the form of dotación over the restitution of lands. Such choices invariably hinged on a wide range of issues specific to each petition and the political and social context under which it was produced. Still, the spatial problematic of restitution (and, conversely, the spatial promise of dota-

111 See Circular no. 19: Instrucciones para identificar tierras solicitadas en restitución, December 25, 1925, AGEV, CLA, caja 4.

112 Quoted in Womack, *Zapata and the Mexican Revolution*, 227.

113 de Gortari to Aurelio Ortega, December 14, 1931, Tonayán, exp. 72; de Gortari to Ing. Jesús Medina, May 19, 1931, exp. Omealca.

ción) is hard to ignore. For many villagers, restitution was historical vindication but for agrarian bureaucrats, its historicity was precisely the problem. Whereas restitution revived a whole gamut of complex questions regarding jurisdiction and property—between as well as within municipalities—and threatened to perpetuate postrevolutionary instability, dotación rescinded that spatial order and supplanted it with an entirely new one. As such, while still nominally accommodating popular demands, it gave the agrarian bureaucracy a significant amount of control over how reform would reshape the spatial and social reality of the countryside. It gave an ascendant political elite, moreover, some measure of control over their political destiny. Finally, with every resolution accompanied by a definitive map, it served to foster the kind of productive and political control over land and people that had so often eluded previous administrations. As John Tutino has observed, if agrarian insurgents fought for *tierra y libertad* (land and liberty), they got *tierra y estado* (land and state).[114]

A last look at Francisco Canóvas's report from Soledad Atzompa is telling. Ordered to resolve a growing conflict between the municipalities of Acultzingo and Soledad Atzompa over lands of the ex-hacienda of Tecamaluca, he suggested that the latter be given a dotación of land from the hacienda due to their poverty. His lengthy letter ended with a revealing conclusion. Dotación, he argued, would not only offer redress to their suffering, it would restore order to the countryside in two complementary ways: by putting an end to conflicts over land and thereby creating an "appropriate environment in which to definitively fix municipal limits."[115]

CONCLUSION

In August 1930 CLA surveyor Manuel Huerta arrived in Acultzingo. The inhabitants of the village, after nearly a decade of (im)patience, had recently written the CLA asking for advice on how to speed up the process of restitution.[116] Huerta came to offer a suggestion: submit a request for

114 Tutino, *From Insurrection to Revolution*, 8. See also Womack, *Zapata and the Mexican Revolution*, 369.

115 Francisco Canóvas to Jefe del Depto. de Comunicaciones y Obras Públicas, Jalapa, July 2, 1932, AGEV, RF, Tierras, Límites, exp. Soledad Atzompa.

116 The original petition for restitution was submitted on March 18, 1921.

dotación. Weary of the delays, the villagers agreed and designated a representative that very day to draw up the petition.[117] Within one month Huerta had submitted a full report to de Gortari, and by June the following year Acultzingo received a provisional grant of 879 hectares. Newly elected President Cárdenas made the grant definitive on April 9, 1934.[118] Only three days later, petitioners from Tonayán received word that their provisional land grant had also been finalized. Having first petitioned for restitution in 1917, they had waited even longer than Acultzingo for movement on their request and by 1929 decided to petition for dotación. Their already slim chances for restitution had vanished along with their primordial titles and historical maps during Adolfo de la Huerta's siege of Xalapa in 1923.[119]

That these communities, like many others, eventually relented in their petitions for restitution is not surprising. Exasperated by the process, they must have feared they would never see revolutionary justice. Confronted with the varied pace and interpretations of reform, it did not take much foresight to realize that continued pursuit of restitution might ensure that no lands were ever returned. The villagers of Coacoatzintla, one of Tonayán's neighbors, solicited for restitution on January 26, 1915, eleven days after Carranza issued his decree. As of October 1979 they were still waiting.[120]

See Pueblo de Acultzingo to Gobernador del Estado, March 18, 1921, exp. Acultzingo.

117 Vecinos de Acultzingo to Comisión Local Agraria de Veracruz, March 15, 1930, exp. Acultzingo, and "En el pueblo de Acultzingo," August 7, 1930, exp. Acultzingo. Huerta did not suggest they change their solicitation but that they simply request a dotación while keeping the request for restitution active.

118 Manuel Huerta to Salvador de Gortari, "Informe sobre los trabajos ejecutados en el Mpio. de Acultzingo, ex-Cantón de Orizaba de este Estado, con motivo de la solicitud de tierras que los vecinos tienen solicitados," September 30, 1930, exp. Acultzingo; "En el pueblo de Acultzingo," June 12, 1931, exp. Acultzingo; and Secretaría de la Reforma Agraria: Perfil.

119 José Gutiérrez, Presidente de la Comisión Local Agraria, to Oficial Mayor de la Comisión Nacional Agraria, January 27, 1926, Tonayán, exp. 72; Salvador de Gortari to Gustavo Lezama, February 7, 1929, Tonayán, exp. 72. The provisional grant was given in 1931. See Ing. Eustolio Delgado to Ing. Presidente de la Comisión Local Agraria, June 12, 1931, Tonayán, exp. 72.

120 Secretaría de la Reforma Agraria: Perfil, 3.

✳

EPILOGUE

"These questions will never end"

This study ends where it began: with Miguel de la Madrid's dream of a national, rural cadastre in 1985, midway through his six-year term in office. The attempt never came to fruition, in part because of the vast distance separating what existed on the ground and what appeared in legal titles, or (in the wonderfully pithy phrase of Luc Cambrezy and Yves Marchal) that space between *el hecho y el derecho* (what is and what is said to be).[1] That such was the result should not have come as a surprise. After all, less than a decade earlier, a similar effort to untangle rights to land and territory in the state of Hidalgo foundered for precisely the same reasons. When land invasions rocked the region of Huejutla in the late 1970s, government officials and agrarian bureaucrats confronted a situation in which the legal distinctions, land categorizations, and forms of tenure they had learned about in law school and government pamphlets simply did not apply.[2] Even local officials and cultivators did not understand the legal status of lands nor the legal categories of landholdings.[3] The federal government managed to put matters to (temporary) rest only by expropriating several million hectares of land to "normalize" the otherwise opaque landscape.[4]

As a strategy to make sense of the productive and political landscape,

1 Cambrezy and Marchal, *Crónicas de un territorio fraccionado*, 134–51. For an excellent examination of these issues in contemporary Michoacán, see Nuijten, *Power, Community, and the State*.
2 See Schryer, "Peasants and the Law."
3 Ibid.
4 Ibid., 305.

normalization became the norm.[5] If the granting of ejidos after the revolution had promised a spatial fix for all the territorial ambiguities that threatened to grind state-builders' plans into the ground, by the 1970s the ejido itself had become as confusing—as fugitive—to the regime as those landscapes of the past. None other than Carlos Salinas de Gortari, de la Madrid's successor at Los Pinos, invoked such opacity when issuing his reform agenda for Article 27 of the Mexican constitution, suggesting that the fixity provided through his proposed reforms would benefit ejidatarios.[6]

The reforms to Article 27, which took effect in February 1992, dictated that ejido land could now be rented, privatized, and sold to national and foreign firms and individuals. This represented a remarkable moment in postrevolutionary Mexican history: the slaughter of one of the sacred cows of revolutionary iconography. In practice, many of the ideals of the revolution had long ago been put out to pasture by Mexico's new postrevolutionary elite, most of them affiliated in one form or another with the ruling Institutional Revolutionary Party. Even so, the revolution's symbols and mythology had always been paid the proper respect in official rhetoric and certain icons—a nationalized oil industry; the ejido—appeared untouchable. Now, even those were open game, although not without significant consequences, as the uprising in Chiapas by a new generation of Zapatistas quickly proved.

The amendments to Article 27 were part of a larger liberalization of the Mexican economy. In accordance with the principles of the North American Free Trade Agreement, ejido land was to be opened up to market circulation. For advocates of neoliberalism, this represented a long-overdue state withdrawal from the agrarian sector and an empowering of peasants to make their own decisions as "legitimate owners of their land."[7] In reality, the reforms seem to have had the opposite effect. As

5 See Murphy, "To Title or Not to Title."

6 See his comments in *La Jornada*, November 15, 1991. As one author has observed, "after more than 70 years of agrarian reform, government authorities still do not know the actual distribution of most of the estimated 4.6 million agricultural parcels and 4.3 million house plots in Mexico or the boundaries of settlement areas and common lands for most of the country's 28,058 ejidos and agrarian communities." Zendejas, "Appropriating Governmental Reforms," 41.

7 The words are de Gortari's, quoted in Stephen, "Accommodation and Resistance."

Lynn Stephen has observed, the notion of empowerment is remarkable given that "very few peasant organizations participated in the design of the certification program nor were they consulted about its implementation."[8] The assumption that the reforms have instituted a state withdrawal from the agrarian sector is equally suspect. In order to rationalize landholdings through surveys, certificates, and titles, a vast new state apparatus and new modes of state penetration have been created. Three new agencies—the Agrarian Attorney General's office, the National Agrarian Registry, and Agrarian Tribunals—now dominate the agrarian bureaucracy, assisted by already existing agencies, such as the National Institute of Statistics, Geography, and Information Management [INEGI]; the Ministry of Agrarian Reform; the Ministry of Agriculture and Hydraulic Resources; and the Ministry of Social Development.[9] Indeed, the certification and titling process, and the concomitant participation of agrarian bureaucrats and surveyors, is remarkably similar to the procedures through which ejidos were originally created after the revolution.

None of this is surprising. Neoliberalism, like its forefather, relies upon substantial state intervention, despite all of its outward disdain for government. The invisible hand of the market has always required the long arm of the state to ensure a long reach and firm grip. Karl Polanyi, more than half a century ago, incisively observed that "the emergence of national markets was in no way the result of the gradual and spontaneous emancipation of the economic sphere from governmental control. On the contrary, the market has been the outcome of a conscious and often violent intervention on the part of government which imposed the market organization on society for noneconomic ends."[10] Among other things, the production of a space—an abstracted and "disembedded" stage—upon which the supposedly self-regulating market could perform its miracles demanded government intervention.

Whether these new spatial fixes will rectify enduring social problems of poverty and political repression, as proponents of the market claim, remains to be seen. If history is any indication, they may well exacerbate them, a fact captured in the ambivalence with which many campesinos

8 Ibid.

9 Pisa, "Popular Response to the Reform of Article 27," 270. See also Nuijten, "Changing Legislation and a New Agrarian Bureaucracy."

10 Polanyi, *The Great Transformation*, 250.

view new surveying and titling procedures. One thing remains certain: the very process of implementing them will not be an easy one. In a story all too familiar now, errors in measurement and mapping by both post- and prerevolutionary surveyors have come back to haunt current attempts at certification and titling. INEGI, the agency responsible for mapping ejido territory and borders, relies upon images that continue to confuse and conflate political territory and agricultural land.[11] Moreover, campesinos refer to missing but definitive ejido maps to make their own claims. Rather than the technologically inspired optimism expressed by the author of PROCEDE's recent pamphlet, who suggested that the combination of "modern surveying equipment and the participation of each and every one of the ejido's inhabitants" would yield precise measurements, the lamentation of Xalapa's prefect more than a century ago appears more apropos: "[T]hese questions will never end."[12]

Still, like contemporary reformers, he proffered technical solutions to what are deeply historical and social questions.[13] As I have suggested, surveying and mapping were never mere technical procedures: they were, and remain, profoundly social and political processes. Regardless of, say, improvements in cartographic technologies or more rational archiving of documents, maps, and titles, struggles over the production and representation of space persisted precisely for this reason. By recognizing and recuperating such struggles, the historically contingent dimensions of space itself can be disentangled from the mirage of its own transparency. It is this, I believe, that gives the histories I have recounted in previous chapters their contemporary salience. A representational image, such as a map, is attractive precisely because its authoritative posture, its overdetermined quality, severs history from its loose ends. As the complexities and contingencies that attended the continual cre-

11 Cambrezy and Marchal, *Crónicas de un territorio fraccionado*, 134, 157. For a particularly telling instance of such confusion, and the dire consequences it could have, see Aubry and Inda, "En los Chimalapas," and Matilde Pérez and Víctor Ruiz Arrazola, "La solución en la región no será decisión unilateral," *La Jornada* August 9, 1999, 48.

12 "La medición del Ejido," pamphlet produced by Programa de Certificación de Derechos Ejidales y Titulación de Solares Urbanos; Jefe Político de Xalapa to Gobernador del Estado, February 7, 1900, exp. Tonayán, ff. 115–19r–v.

13 For a recent examination that looks at how the current reforms to Article 27 seek to reduce complicated land rights to technical problems, see Nuijten, "Family Property and the Limits of Intervention."

ation and recreation of space have faded with time, so too have the words and worlds that lay behind the facades that resulted from them. We are left with fixed images and fixed histories, with stages and teleologies. "Fixations," wrote Michel de Certeau, "constitute procedures for forgetting. The trace left behind is substituted for practice. It exhibits the (voracious) property that the geographical system has of being able to transform action into legibility, but in doing so it causes a way of being in the world to be forgotten."[14] The power of fixations is their ability to disempower; it is their ability to rewrite the place of history as the space of legitimacy and to convert contingency into inevitability. Indeed, the ultimate projection of the stage space is the twin suggestion that a specific social and spatial order is natural and the future, hence, inevitable.[15] Perhaps dismantling the fixed traces of the stage space, by emphasizing the historical and social processes that conditioned its creation, may constitute a procedure to recuperate and imagine other possibilities, other ways of being in the world, and other opportunities that were figuratively and literally foreclosed.

14 de Certeau, *The Practice of Everyday Life*, 97.

15 For a recent but already "classic" neoconservative claim to this effect, see Fukuyama, *The End of History and the Last Man*.

BIBLIOGRAPHY

ARCHIVES

Mexico City
Archivo General Agrario
Archivo General de la Nación
Archivo Histórico del Agua
Archivo Particular Daniela Canóvas Rebling
      Fondo Francisco Canóvas y Pasquel
Biblioteca Nacional, Universidad Nacional Autónoma de México
      Colección La Fragua
Centro de Estudios de Historia Mexicana
      Colección Félix Díaz
      Colección Venustiano Carranza
Mapoteca Manuel Orozco y Berra
      Archivo de la Comisión Geográfico-Exploradora
      Colección General
Sociedad Mexicana de Geografía y Estadística
      Archivo Histórico
Universidad Iberoamericana
      Colección Porfirio Díaz

Veracruz
Archivo General del Estado de Veracruz
Archivo Municipal de Acultzingo
Archivo Municipal de Misantla
Archivo Municipal de Orizaba

United States
Benson Latin American Collection at the University of Texas, Austin
      Génaro García Collection
      Vicente Riva Palacio Papers

WORKS CITED

Aboites, Luis. *El agua de la nación: Una historia política.* Mexico City: CIESAS, 1997.

Aboites, Luis, and Alba Morales Cosme. "Amecameca, 1922: Ensayo sobre centralización política y estado nacional en México." *Historia Mexicana* 193 (1999), 55–93.

Abrams, Philip. "Notes on the Difficulty of Studying the State." *Journal of Historical Sociology* 1, no. 1 (1988), 58–89.

Acuña, René, ed. *Relaciones geográficas del siglo XVI.* 10 vols. Mexico City: UNAM, 1981–1988.

Adorno, Rolena. "Reconsidering Colonial Discourse for Sixteenth- and Seventeenth-Century Spanish America." *Latin American Research Review* 28, no. 3 (1993), 135–45.

Agnew, Jean-Christophe. *Worlds Apart: The Market and the Theater in Anglo-American Thought, 1550–1750.* Cambridge: Cambridge University Press, 1986.

Aguilar Camín, Héctor, and Lorenzo Meyer. *In the Shadow of the Mexican Revolution: Contemporary Mexican History, 1910–1989.* Translated by Luis Alberto Fierro. Austin: University of Texas Press, 1993.

Aguilar Robledo, Miguel. "Land Use, Land Tenure, and Environmental Change in the Jurisdiction of Santiago de los Valles de Oxitipa, Eastern New Spain, Sixteenth to Eighteenth Centuries." Ph.D. dissertation, University of Texas at Austin, 1999.

Ajofrín, Francisco de. "Diario del Viaje que hicimos a Mexico, Fray Francisco de Ajofrín y Fray Fermín de Olite, Capuchinos." In Poblett Miranda, ed., *Cien Viajeros en Veracruz*, vol. 2.

Althusser, Louis, and Etienne Balibar. *Reading Capital.* New York: Pantheon Books, 1971.

Alvarado, Julio. *Anexo de Memoria de la Secretaría de Fomento, Comisión Geográfico-Exploradora, 1892–96.* Mexico City: Imprenta de la Secretaría de Fomento, 1897.

———. *Comisión Geográfico-Exploradora. Catálogo de los objetos que componen el contingente de la expresada Comisión, precedido de una reseña abreviada sobre su organización y trabajos. Exposición Universal Internacional de París en 1900.* Mexico City: Imprenta de la Secretaría de Fomento, 1900.

———. *The Geographic and Exploring Commission of Mexico. Sketch of its Organization and Labors.* Buffalo: Gies, n.d.

———. "Informe de la Comisión Geográfico-Exploradora." *Memoria presentada al Congreso de la Unión . . . 1897 a 1900.*

Anales de la Asociación de Ingenieros y Arquitectos de México. Mexico City: Oficina Tip. de la Secretaría de Fomento, 1886.

Anderson, Perry. *Lineages of the Absolutist State.* London: Verso Press, 1979.

Anguiano, Angel. "Cartografía Mexicana." *Boletín de la Sociedad Mexicana de Geografía y Estadística,* 5a época, tomo 7 (1914), 168–92.

Anna, Timothy. *Forging Mexico, 1821–1835.* Lincoln: University of Nebraska Press, 1998.

Appelbaum, Nancy P. *Muddied Waters: Race, Region and Local History in Colombia, 1846–1948.* Durham: Duke University Press, 2003.

Archer, Christon. "Discord, Disjunction, and Reveries of Past and Future Glories: Mexico's First Decades of Independence, 1810–1853." *Mexican Studies/Estudios Mexicanos* 16, no. 1 (2000), 189–210.

Arroníz, Joaquín. *Ensayo de una historia de Orizaba.* Mexico City: Editorial Citlateptl, 1980 [1867].

Aubry, Andrés, and Angélica Inda. "En los Chimalapas, enredos cartográficos y legislativos." *La Jornada,* August 9, 1999, 48.

Aznar Barbachano, Tomás. "Importancia del estudio de la geografía y estadística como base fundamental de un buen gobierno." *Boletín de la Sociedad Mexicana de Geografía y Estadística,* 1a. época, tomo 8 (1860), 460–63.

Bachelard, Gaston. *The Poetics of Space.* Translated by Maria Jolas. New York: Orion Press, 1964.

Bakhtin, Mikhail. *The Dialogic Imagination: Four Essays by M. M. Bakhtin.* Edited by Michael Holquist. Austin: University of Texas Press, 1981.

Barnes, Trevor, and Derek Gregory. *Reading Human Geography: The Poetics and Politics of Inquiry.* London: John Wiley and Sons, 1997.

Barthes, Roland. *Mythologies.* Translated by A. Lavers. New York: Hill and Wang, 1986.

Basso, Keith. *Wisdom Sits in Places: Landscape and Language among the Western Apache.* Albuquerque: University of New Mexico Press, 1995.

Bazant, Mílada. "La enseñanza agrícola en México: Prioridad gubermental e indiferencia social (1853–1910)." *Historia Mexicana* 32, no. 3 (1983), 349–88.

Beezley, William H., Cheryl English Martin, and William E. French, eds. *Rituals of Rule, Rituals of Resistance: Public Celebrations and Popular Culture in Mexico.* Wilmington: SR Books, 1994.

Belyea, Barbara. "Images of Power: Derrida/Foucault/Harley." *Cartographica* 29, no. 2 (1992), 1–9.

Benjamin, Thomas. *La Revolución: Mexico's Great Revolution as Memory, Myth and History.* Austin: University of Texas Press, 2000.

Berger, John. *Into Their Labours: A Trilogy (Pig Earth, Once in Europe, Lilac and Flag).* New York: Pantheon Books, 1990.

———. *Ways of Seeing.* London: Penguin Books, 1972.

Bernstein, Marvin D. *The Mexican Mining Industry, 1890–1950: A Study of the Interaction of Politics, Economics, and Technology.* Albany: State University of New York, 1964.

Bhabha, Homi. "DissemiNation: Time, Narrative and the Margins of the Modern Nation." In *The Location of Culture*. London: Routledge, 1994.

Bhabha, Homi, ed. *Nation and Narration*. London: Routledge, 1990.

Blázquez Domínguez, Carmen, ed. *Estado de Veracruz: Informes de sus Gobernadores, 1826–1986*. Xalapa: Gobierno del Estado de Veracruz, 1986.

Boelhower, William. "Inventing America: A Model of Cartographic Semiosis." *Word and Image* 4, no. 2 (1988), 475–97.

Borah, Woodrow. *Justice by Insurance: The General Indian Court of Colonial Mexico*. Berkeley: University of California Press, 1983.

Bourdieu, Pierre. *Outline of a Theory of Practice*. Translated by Richard Nice. Cambridge: Cambridge University Press, 1977.

Brading, D. A. *The First America: The Spanish Monarchy, Creole Patriots, and the Liberal State, 1492–1867*. Cambridge: Cambridge University Press, 1991.

Brand, Dana. *The Spectator and the City in Nineteenth-Century American Literature*. Cambridge: Cambridge University Press, 1991.

Braudel, Fernand. *La Méditerranée et le monde méditerranéen à l'époque de Philippe II*. 2d edition. Paris: Libraire Armand Colin, 1966.

Buelna, Eustaquio. "Peregrinación de los Aztecas y nombres geográficos indígenas de Sinaloa." *Boletín de la Sociedad Mexicana de Geografía y Estadística* 4a. época, tomo 2 (1890), 315–462.

Burchell, Graham, Colin Gordon, and Peter Miller, eds. *The Foucault Effect: Studies in Governmentality*. Hemel Hempstead: Harvester Wheatsheaf, 1991.

Burnett, D. Graham. *Masters of All They Surveyed: Exploration, Geography and a British El Dorado*. Chicago: University of Chicago Press, 2000.

Caballero, Manuel. *Primer almanaque histórico, artístico y monumental de la República Mexicana, 1883–1884*. Mexico City and New York: Chas. M. Green Printing, 1883.

Cabrera, Luis. *Diccionario de Aztequismos*. Mexico City: Ediciones Oasis, 1974.

Cambrezy, Luc, and Yves Marchal. *Crónicas de un territorio fraccionado: De la hacienda al ejido (centro de Veracruz)*. Mexico City: Ediciones Larousse, 1992.

Capel, Horacio. "Institutionalization of Geography and Strategies of Change." In *Geography, Ideology and Social Concern*, edited by David R. Stoddart. Oxford: Blackwell, 1981.

Capilla, Alberto. "Los yacimientos de fierro de Tatatila, Cantón de Jalapa, E. de Veracruz." *Boletín de la Secretaría de Fomento: Número de Propaganda* (July 1905), folleto 12: 1–8. Mexico City: Imprenta de la Secretaría de Fomento, 1905.

Cardoso, Ciro, ed. *México en el siglo XIX: Historia económica y de la estructura social*. Mexico City: Editorial Nueva Imagen, 1980.

Carreno, D. Alberto M. "La evolución económica de la raza indígena." *Boletín de la Sociedad Mexicana de Geografía y Estadística* 5a. época, tomo 5 (1912), 59–76.

Carson, W. E. *Mexico: The Wonderland of the South*. New York: Macmillan, 1909.

Carter, Erica, James Donald, and Judith Squires, eds. *Space and Place: Theories of Identity and Location*. London: Lawrence and Wishart, 1993.

Carter, Paul. *The Lie of the Land*. London: Faber and Faber, 1996.

———. *The Road to Botany Bay: An Exploration in Landscape and History*. Chicago: University of Chicago Press, 1987.

Casey, Edward S. *The Fate of Place: A Philosophical History*. Berkeley: University of California Press, 1997.

Chenaut, Victoria. *Aquellos que vuelan: Los totonacos en el siglo XIX*. Mexico City: CIESAS, 1995.

———. "Fin de siglo en la costa totonaca: Rebeliones indias y violencia regional, 1891–96." In Chenaut, ed., *Procesos rurales e historia regional*.

———, ed. *Procesos rurales e historia regional (sierra y costa totonacas de Veracruz)*. Mexico City: CIESAS, 1996.

Clarke, G. N. G. "Taking Possession: The Cartouche as Cultural Text in Eighteenth Century American Maps," *Word and Image* 4, no. 2 (1988), 455–74.

Collado, María del Carmen. "Antonio García Cubas." In Ortega y Medina and Camelo, eds., *En busca de un discurso integrador*.

Cohn, Bernard. *Colonialism and Its Forms of Knowledge*. Chicago: University of Chicago Press, 1996.

Comaroff, Jean, and John Comaroff. *Of Revelation and Revolution*. Vol. 1, *Christianity, Colonialism and Consciousness in South Africa*. Chicago: University of Chicago Press, 1991.

———. *Of Revelation and Revolution*. Vol. 2, *The Dialectics of Modernity on a South African Frontier*. Chicago: University of Chicago Press, 1997.

Comisión Nacional Agraria. *Leyes y disposiciones referentes a restituciones y dotaciones de tierras para ejidos*. Mexico City: Imprenta de la Dirección de Estudios Geográficos y Climatológicos, 1922.

———. *Recopilación Agraria*. Mexico City: Imprenta de la Dirección de Estudios Geográficos y Climatológicos, 1924.

Conkling, Alfred R. *Appleton's Guide to Mexico*. New York: D. Appleton, 1884.

Córdova, Arnaldo. *La ideología de la revolución mexicana: La formación del nuevo régimen*. Mexico City: Ediciones Era, 1973.

Coronil, Fernando. *The Magical State: Nature, Money and Modernity in Venezuela*. Chicago: University of Chicago Press, 1997.

Corrigan, Philip, and Derek Sayer. *The Great Arch: English State Formation as Cultural Revolution*. Oxford: Blackwell, 1991.

Cosgrove, Denis. "Prospect, Perspective and the Landscape Idea." *Transactions of the Institute of British Geographers* 10 (1985), 45–62.

Craib, Raymond B. "Cartography and Power in the Conquest and Creation of New Spain." *Latin American Research Review* 35, no. 1 (2000), 7–37.

————. "A Nationalist Metaphysics: State Fixations, National Maps, and the Geo-Historical Imagination in Nineteenth-Century Mexico," *Hispanic American Historical Review* 82, no. 1 (February 2002), 33–68.

————. "State Fixations, Fugitive Landscapes: Mapping, Surveying and the Spatial Creation of Modern Mexico, 1850–1930." Ph.D. dissertation, Yale University, 2001.

Craib, Raymond B., and D. Graham Burnett. "Insular Visions: Cartographic Imagery and the Spanish American War." *The Historian* 61, no. 1 (fall 1998), 101–18.

Craig, Ann L. *The First Agraristas: An Oral History of a Mexican Agrarian Reform Movement.* Berkeley: University of California Press, 1983.

Cronon, William. *Changes in the Land: Indians, Colonists and the Ecology of New England.* New York: Hill and Wang, 1983.

Cuadros Caldas, Julio. *Catecismo Agrario.* Mexico City: CIESAS-RAN, 1999 [1923].

————. *Mexico-Soviet.* Puebla: Santiago Loyo, 1926.

Dabdoub, Claudio. *Historia de El Valle del Yaqui.* Mexico City: Librería de Manuel Porrua, 1964.

Daston, Lorraine, and Peter Galison. "The Image of Objectivity." *Representations* 40 (autumn 1992), 81–128.

de Certeau, Michel. *The Practice of Everyday Life.* Translated by Steven Rendall. Berkeley: University of California Press, 1984.

de P. Piña, Francisco. "La Comisión Geográfico-Exploradora y la influencia de sus trabajos en la geografía del país." *Boletín de la Sociedad Mexicana de Geografía y Estadística,* 5a. época, tomo 3 (1908), 281–97.

————. "Importancia de los trabajos geográficos e históricos del Señor Ingeniero don Antonio García Cubas." *Boletín de la Sociedad Mexicana de Geografía y Estadística,* 5a. época, tomo 3 (1908), 389–409.

de Vos, Jan. *Las fronteras de la frontera sur: Reseña de los proyectos de expansión que figuraron la frontera entre México y Centroamérica.* Villahermosa: Universidad Juárez Autónoma de Tabasco, CIESAS, 1993.

Deane, Seamus. Introduction to *Nationalism, Colonialism and Literature,* by Terry Eagleton, Fredric Jameson, and Edward Said. Minneapolis: University of Minnesota Press, 1990.

Deleuze, Gilles, and Felix Guattari. *A Thousand Plateaus: Capitalism and Schizophrenia.* Minneapolis: University of Minnesota Press, 1987.

Dening, Greg. *Islands and Beaches: Discourse on a Silent Land: Marquesas, 1774–1880.* Honolulu: University of Hawaii Press, 1980.

————. "A Poetic for Histories: Transformations That Present the Past." In *Clio in Oceania: Towards a Historical Anthropology,* edited by Aletta Biersack. Washington, D.C.: Smithsonian Institution Press, 1991.

Díaz, Agustín. *Exposición internacional colombina de Chicago en 1893. Catálogo de los objetos que componen el contingente de la Comisión, precedido de algunas notas sobre su organización y trabajos.* Xalapa-Enríquez: Tipografía de la Comisión Geográfico-Exploradora, 1893.

———. "Informe de la Comisión Geográfico-Exploradora." *Memoria presentada al Congreso de la Unión . . . enero de 1883 a junio de 1885,* vol. 1.

———. "Informe sobre el estado actual de la cartografía." *Memoria presentada al Congreso de la Unión . . . 1877.*

———. *Memoria de la Comisión Geográfico-Exploradora presentada al oficial mayor, encargada de la Secretaría de Fomento, sobre los trabajos ejecutados durante el año fiscal de 1878 a 1879.* Mexico City: Imprenta de Francisco Díaz de León, 1880.

Díaz, Porfirio. *Informe que en el último día de su período constitucional da a sus compatriotas el presidente de los Estados Unidos Mexicanos, Porfirio Díaz, acerca de los actos de su administración.* Mexico City: Tipografía de Gonzalo A. Estevan, 1880.

Díaz del Castillo, Bernal. *The Conquest of New Spain.* Translated and with an introduction by J. M. Cohen. London: Penguin Books, 1963.

Díaz Rivero, Francisco. *Estudio preliminar sobre la manera de proceder al levantamiento de la carta militar, catastral, civil y política del país.* Reproducción facsimilar del manuscrito de 1896. Mexico City: Taller de Offset de la Comisión Nacional de Irrigación, 1946.

"Dictamen presentado a la Sociedad de Geografía y Estadística por la comisión especial que subscribe con objeto de pedir al supremo gobierno que declare propiedad nacional los monumentos arqueológicos de la república." *Boletín de la Sociedad Mexicana de Geografía y Estadística,* 1a. época, tomo 8 (1860), 438–442.

Dirlik, Arif. *The Postcolonial Aura: Third World Criticism in the Age of Global Capitalism.* Boulder: Westview Press, 1997.

Duara, Prasenjit. *Rescuing History from the Nation: Questioning Narratives of Modern China.* Chicago: University of Chicago Press, 1995.

Ducey, Michael T. "Indios liberales y liberales indigenistas: Ideología y poder en los municipios rurales de Veracruz, 1821–1890." Paper presented at the Latin American Studies Association meeting, 2000.

———. "Indios liberales y tradicionales: Tres casos de privatización de la tierra comunal." *Memorial: Boletín del Archivo General del Estado de Veracruz* 2, no. 4 (1999), 13–18.

———. "Liberal Theory and Peasant Practice: Land and Power in Northern Veracruz, Mexico, 1826–1900." In *Liberals, the Church, and Indian Peasants: Corporate Lands and the Challenge of Reform in Nineteenth-Century Spanish America,* edited by Robert H. Jackson. Albuquerque: University of New Mexico Press, 1996.

———. "Tierras comunales y rebeliones en el norte de Veracruz antes del Porfiriato, 1821–1878: El proyecto liberal frustrado." *Anuario: Universidad Veracruzana* 6 (1989), 209–29.

———. "Viven sin ley ni rey: Rebeliones coloniales en Papantla, 1760–1790."
In Chenaut, ed. *Procesos rurales e historia regional.*

Duclos Salinas, Adolfo. *The Riches of Mexico and Its Institutions. Edition for the World's Fair Exposition.* St. Louis: Nixon-Jones Printing, 1893.

Eagleton, Terry. *Literary Theory: An Introduction.* Minneapolis: University of Minnesota Press, 1983.

Edgerton Jr., Samuel Y. *The Renaissance Rediscovery of Linear Perspective.* New York: Basic Books, 1975.

Edney, Matthew. "Cartographic Culture and Nationalism in the Early United States: Benjamin Vaughan and the Choice for a Prime Meridian, 1811." *Journal of Historical Geography* 20, no. 4 (1994), 384–95.

———. *Mapping an Empire: The Geographical Construction of British India, 1765–1843.* Chicago: University of Chicago Press, 1997.

Escalante Gonzalbo, Fernando. *Ciudadanos imaginarios: Memorial de los afanes y desventuras de la virtud y apología del vicio triunfante en la república mexicana.* Mexico City: El Colegio de México, 1992.

Escobar Ohmstede, Antonio, and Jacqueline Gordillo. "Defensa o Despojo? Territorialidad indígena en las Huastecas, 1856–1930." In Escobar Ohmstede et al., *Estudios campesinos.*

Escobar Ohmstede, Antonio, Jacqueline Gordillo, María Rosa Gudiño, Guillermo Palacios, Gabriela Acosta, Arnulfo Embriz, and Juan Matamala. *Estudios campesinos en el Archivo General Agrario.* Mexico City: CIESAS-RAN, 1998.

Escolar, Marcelo. "Exploration, Cartography, and the Modernization of State Power." *International Social Science Journal* 151 (May 1997), 55–76.

Estado de San Luis Potosí. *Carta General del Estado de San Luis Potosí, levantado por iniciativa de su actual gobernador Gral. Carlos Diez Gutiérrez, por la Comisión Geográfico-Exploradora.* Mexico City: n.p., 1894.

Estado de Veracruz. *Colección de leyes, decretos y circulares del Estado de Veracruz Llave, año de 1871.* Jalapa: Imprenta Veracruzana, 1893.

———. *Colección de leyes . . . 1876–78.* Xalapa-Enríquez: Imprenta del Gobierno del Estado, 1894.

———. *Colección de leyes . . . 1879.* Orizaba: Imprenta del Gobierno del Estado, 1885.

———. *Colección de leyes . . . 1881.* Jalapa: Imprenta del Gobierno del Estado de Veracruz, 1885.

———. *Colección de leyes . . . 1886.* Xalapa-Enríquez: Imprenta del Gobierno del Estado de Veracruz, 1899.

———. *Colección de leyes . . . 1887.* Xalapa-Enríquez: Tipografía del Gobierno del Estado, 1899.

———. *Colección de leyes . . . 1889.* Jalapa: Tipografía del Gobierno del Estado, 1889.

———. *Colección de leyes*... *1890*. Xalapa-Enríquez: Tipografía del Gobierno del Estado, 1890

———. *Colección de leyes*... *1891*. Xalapa-Enríquez: Tipografía del Gobierno del Estado, 1891.

———. *Colección de leyes*... *1895*. Xalapa-Enríquez: Tipografía del Gobierno del Estado, 1896.

———. *Colección de leyes*... *1898*. Xalapa-Enríquez: Tipografía del Gobierno del Estado, 1898.

———. *Colección de leyes*... *1899*. Xalapa-Enríquez: Tipografía del Gobierno del Estado, 1899.

———. *Colección de leyes*... *1901*. Xalapa-Enríquez: Tipografía del Gobierno del Estado, 1901.

Fabian, Johannes. *Time and the Other: How Anthropology Makes Its Object*. New York: Columbia University Press, 1983.

Falcón, Romana. "Force and the Search for Consent: The Role of the Jefaturas Políticas of Coahuila in National State Formation." In Joseph and Nugent, eds., *Everyday Forms of State Formation*.

Falcón, Romana, and Raymond Buve, eds. *Don Porfirio President*..., *nunca omnipotente: Hallazgos, reflexiones y debates, 1876–1911*. Mexico City: Universidad Iberoamericana, 1998.

Fanon, Frantz. *The Wretched of the Earth*. Translated by Constance Farrington. New York: Grove, 1963.

Faragher, John Mack. *Sugar Creek: Life on the Illinois Prairie*. New Haven: Yale University Press, 1986.

Feld, Steven, and Keith H. Basso, eds. *Senses of Place*. Santa Fe: School of American Research, 1994.

Fenochio, Juan. *Informe acerca del Mineral de Zomelahuacán, presentada a la Secretaría de Fomento*. Mexico City: Secretaría de Fomento, 1883.

Ferguson, James. *The Anti-Politics Machine: "Development," Depoliticization, and Bureaucratic Power in Lesotho*. Minneapolis: University of Minnesota Press, 1994.

Ferrari Pérez, Fernando. "Informe de los trabajos de la Comisión Geográfico-Exploradora." In *Memoria presentada al Congreso de la Unión*... *de 10 de enero de 1901 al 31 de diciembre de 1904*.

Feyerabend, Paul. *Against Method*. London: New Left Books, 1975.

Figueroa Domenech, J. *Guía general descriptiva de la República Mexicana*. Barcelona: Imprenta de Henrich y Compañía, 1899.

Florescano Mayet, Sergio. "El proceso de destrucción de la propiedad comunal de la tierra y las rebeliones indígenas en Veracruz, 1826–1910." *La palabra y el hombre* 52 (October–December 1984), 5–18.

Foucault, Michel. *The Archaeology of Knowledge*. Translated by A. M. Sheridan Smith. New York: Pantheon Books, 1972.

———. *The Order of Things: An Archaeology of the Human Sciences.* New York: Vintage Books, 1973.

———. "Questions of Geography." In *Power/Knowledge: Selected Interviews and Other Writings, 1972–1977*, edited by Colin Gordon. New York: Pantheon Books, 1980.

Fowler-Salamini, Heather. *Agrarian Radicalism in Veracruz, 1920–1938.* Lincoln: University of Nebraska Press, 1971.

———. "Tamaulipas: Land Reform and the State." In *Provinces of the Revolution: Essays on Regional Mexican History, 1910–1929*, edited by Thomas Benjamin and Mark Wasserman. Albuquerque: University of New Mexico Press, 1990.

Franck, Harry A. *Trailing Cortez through Mexico.* New York: Frederick A. Stokes, 1935.

Fritszche, Peter. *Reading Berlin, 1900.* Cambridge: Harvard University Press, 1996.

Fukuyama, Francis. *The End of History and the Last Man.* New York: Avon Books, 1993.

Galván Rivera, Mariano. *Ordenanzas de tierras y aguas, o sea formulario geométrico-judicial, para la designación, establecimiento, mensura, amojonamiento y deslinde de las poblaciones, y todas suertes de tierras, sitios, caballerías y criaderos de ganados mayores y menores, y mercedes de aguas, recopiladas a beneficio y obsequio de los pobladores, ganaderos, labradores, dueños, arrendatarios y administradores de haciendas, y toda clase de predios rústicos, de las muchas y dispersas resoluciones dictadas sobre la materia, y vigentes hasta el día en la República Mexicana.* Mexico City: Imprenta de Vicente G. Torres, 1842.

Gama, Valentín. "Consideraciones acerca de la cartografía en México y sobre la manera de promover el adelanto de la misma." In *Primer centenario de la Sociedad Mexicana de Geografía y Estadística, 1833–1933.* Mexico City: Sociedad Mexicana de Geografía y Estadística, 1933.

García Cubas, Antonio. *Atlas geográfico, estadístico e histórico de la República Mexicana.* Mexico City: Imprenta de José Mariano Fernández de Lara, 1858.

———. *Atlas mexicano geográfico y estadístico de los Estados Unidos Mexicanos por Antonio García Cubas.* Mexico City: Debray Sucesores, 1886.

———. *Atlas pintoresco e histórico de los Estados Unidos Mexicanos.* Mexico City: Debray Sucesores, 1885.

———. *Carta general de la República Mexicana. Formado en vista de los datos más recientes y exactos que se han reunido con tal objeto, y constan en la noticia presentada al Exmo. Sr. Ministro de Fomento.* Mexico City: Imprenta de José Mariano Fernández de Lara, 1858.

———. *Carta general en major escala.* In García Cubas, *Atlas geográfico, estadístico e histórico de la República Mexicana.*

———. *Cuadro geográfico, estadístico, descriptivo, e histórico de los Estados Unidos*

Mexicanos. Mexico City: Oficina Tipográfica de la Secretaría de Fomento, 1884.

————. *Diccionario geográfico, histórico y biográfico de los Estados Unidos Mexicanos.* 5 vols. Mexico City: Antigua Imprenta de Murguia, 1888.

————. *Etude géographique, statistique, descriptive et historique des Etats Unis Mexicains.* Mexico City: Imprimerie du Ministère des Travaux Publics, 1889.

————. "Informe sobre el estado actual de la Estadística nacional." *Memoria presentado al Congreso de la Unión . . . 1877.*

————. *El libro de mis recuerdos: Narraciones históricas, anecdóticas y de costumbres Mexicanas anteriores al actual órden social.* Mexico City: Imprenta de Arturo García Cubas Sucesores Hermanos, 1905.

————. *The Republic of México in 1876.* Translated by George E. Henderson. Mexico City: La Enseñanza, 1876.

————. *Memoria para servir a la carta general de la República Mexicana.* Mexico City: Imprenta de Andrade y Escalante, 1861.

————. *Mexico: Its Trade, Industry and Resources.* Translated by William Thompson. Mexico City: Oficina Tipografíca de la Secretaría de Fomento, 1893.

García Cubas, Antonio, Francisco Díaz Covarrubias, and Manuel Fernández. "Dictamen sobre los inconvenientes de variar los nombres a los lugares de la República." *Boletín de la Sociedad Mexicana de Geografía y Estadística,* 2a. época, tomo 1 (1869), 601–4.

García Granados, Ricardo. "Discurso pronunciado por el Socio Ingeniero Ricardo García Granados al entregar a la Comisión Geográfico-Exploradora, el diploma acordado por la Sociedad en memoria del señor Coronel de EME Ingeniero D. Agustín Díaz." *Boletín de la Sociedad Mexicana de Geografía y Estadística,* 5a. época, tomo 3 (1908), 309–12.

García Granados, Ricardo, Francisco de P. Piña, and Ricardo Ortega y Pérez Gallardo. "Dictamen presentado a la Sociedad por los socios ingenieros Ricardo García Granados y Francisco de P. Piña y Ricardo Ortega y Pérez Gallardo, sobre la proposición que hizo la 'Sociedad Cultural, Intelectual, Moral y Física,' de Guadalajara, para cambiar los nombres de los dos picos del volcán de Colima, conocidos por Volcán de Nieve el uno, y el otro por Volcán de Fuego." *Boletín de la Sociedad Mexicana de Geografía y Estadística,* 5a. época, tomo 3 (1908), 461–67.

García Martínez, Bernardo. "La Comisión Geográfico-Exploradora." *Historia Mexicana* 96 (April–June 1975), 485–555.

García Morales, Soledad, and José Velasco Toro, eds. *Memorias e informes de jefes políticos y autoridades del régimen Porfirista, 1883–1911: Estado de Veracruz.* Xalapa: Universidad Veracruzana, 1997.

Gibson, Charles. *The Aztecs under Spanish Rule: A History of the Indians of the Valley of Mexico, 1519–1810.* Stanford: Stanford University Press, 1964.

Glacken, Clarence J. *Traces on the Rhodian Shore: Nature and Culture in Western Thought from Ancient Times to the End of the Eighteenth Century*. Berkeley: University of California Press, 1967.

Godlewska, Anne, and Neil Smith, eds. *Geography and Empire*. Oxford: Blackwell, 1994.

Godoy, José Francisco. *Porfirio Díaz, President of Mexico: The Master Builder of a Great Commonwealth*. New York: G. P. Putnam's Sons, 1910.

Gómez, Macedonio. "Concursos científicos." *Boletín de la Sociedad Mexicana de Geografía y Estadística*, 4a. época, tomo 2 (1894), 500–74.

Gómez, Marte R. *Historia de la Comisión Nacional Agraria*. Mexico City: Centro de Investigaciones Agrarias, 1975.

———, ed. *Primera convención de la Liga de comunidades agrarias y sindicatos campesinos del estado de Tamaulipas*. Mexico City: Editorial Cultura, 1926.

———, ed. *Segunda convención de la Liga de comunidades agrarias y sindicatos campesinos del estado de Tamaulipas*. Mexico City: Editorial Cultura, 1927.

———, ed. *Tercera convención de la Liga de comunidades agrarias y sindicatos campesinos del estado de Tamaulipas*. Mexico City: Editorial Cultura, 1928.

González de la Lama, René. "Los papeles de Díaz Manfort: Una revuelta popular en Misantla (Veracruz), 1885–86." *Historia Mexicana* 31, no. 2 (October–December 1989), 475–521.

González Echevarría, Roberto. *Myth and Archive: A Theory of Latin American Narrative*. Durham: Duke University Press, 1998.

Gray, Albert Zabriskie. *Mexico as It Is, Being Notes of a Recent Tour in That Country*. New York: E. P. Dutton, 1878.

Gregory, Derek. *Geographical Imaginations*. Oxford: Blackwell, 1994.

Gruzinski, Serge. *The Conquest of Mexico: The Incorporation of Indian Societies into the Western World, Sixteenth to Eighteenth Centuries*. Translated by Eileen Corrigan. Cambridge: Polity Press, 1993.

Guardino, Peter. *Peasants, Politics and the Formation of Mexico's National State: Guerrero, 1800–1857*. Stanford: Stanford University Press, 1996.

Gudiño, María Rosa, and Guillermo Palacios. "Peticiones de tierras y estrategias discursivas campesinas: procesos, contenidos y problemas metodológicos." In Escobar Ohmstede et al., *Estudios campesinos*.

Guerra, François-Xavier. *México: Del antiguo régimen a la revolución*. 2 vols. Mexico City: Fondo de Cultural Económica, 1988.

Gupta, Akhil. "Blurred Boundaries: The Discourse of Corruption, the Culture of Politics, and the Imagined State." *American Ethnologist* 22, no. 2 (1995), 375–402.

Gupta, Akhil, and James Ferguson, eds. *Culture, Power, Place: Explorations in Critical Anthropology*. Durham: Duke University Press, 1997.

Hale, Charles. *Liberalism in the Age of Mora*. New Haven: Yale University Press, 1968.

————. *The Transformation of Liberalism in Late Nineteenth-Century Mexico.* Princeton: Princeton University Press, 1989.

Hall, Dawn, ed. *Drawing the Borderline: Artist-Explorers of the U.S.-Mexico Boundary Survey.* Albuquerque: Albuquerque Museum, 1996.

Hamilton, Nora. *The Limits of State Autonomy: Post-Revolutionary Mexico.* Princeton: Princeton University Press, 1982.

Hardin, Garrett. "The Tragedy of the Commons." *Science* 162 (1968), 1243–48.

Harley, J. B. "The Map and the Development of the History of Cartography." In *The History of Cartography*, vol. 1, *Cartography in Prehistoric, Ancient, and Medieval Europe and the Mediterranean*, edited by J. B. Harley and David Woodward. Chicago: University of Chicago Press, 1987.

————. *The New Nature of Maps: Essays in the History of Cartography.* Edited by Paul Laxton. Baltimore: Johns Hopkins University Press, 2001.

————. "Rereading the Maps of the Columbian Encounter." *Annals of the Association of American Geographers* 82, no. 3 (1992), 543–65.

Harvey, David. "Between Space and Time: Reflections on the Geographical Imagination." *Annals of the Association of American Geographers* 80, no. 3 (1990), 418–34.

————. *The Condition of Postmodernity: An Inquiry into the Origins of Cultural Change.* Oxford: Blackwell, 1989.

————. *Social Justice and the City.* London: Edward Arnold, 1973.

Henderson, Timothy J. *The Worm in the Wheat: Rosalie Evans and the Agrarian Struggle in the Puebla-Tlaxcala Valley of Mexico, 1906–1927.* Durham: Duke University Press, 1998.

Henderson, Peter V. N. *Félix Díaz, the Porfirians and the Mexican Revolution.* Lincoln: University of Nebraska Press, 1981.

Hernández Chávez, Alicia. *Anenecuilco: Memoria y vida de un pueblo.* Mexico City: Colegio de Mexico, 1991.

Hernández Fujigaki, Gloria. "Los agrónomos frente a los retos nacionales, 1921–1989." In *Agricultura y agronomía en México: Origen, desarrollo y actualidad. Memoria del primer simposium sobre origen, desarrollo y actualidad de la agricultura y agronomía en México.* Mexico City: Universidad Autónoma Chapingo, 1991.

Hernández Silva, Héctor Cuauhtémoc. *Insurgencia y autonomía: Historia de los pueblos yaquis, 1821–1910.* Mexico City: CIESAS, 1995.

Hillis, Ken. "The Power of the Disembodied Imagination: Perspective's Role in Cartography." *Cartographica* 31, no. 3 (1994), 1–17.

Hobsbawm, E. J. *The Age of Capital, 1848–1875.* New York: Vintage Books, 1996.

————. *Nations and Nationalism since 1870: Programme, Myth and Reality.* Cambridge: Cambridge University Press, 1990.

Hoffman, Odile, and Fernando I. Salmerón Castro, eds. *Nueve estudios sobre el espacio: Representación y formas de apropiación.* Mexico City: CIESAS, 1997.

Holden, Robert. *Mexico and the Survey of Public Lands: The Management of Modernization, 1876–1911*. DeKalb: Northern Illinois University Press, 1994.

Hu-DeHart, Evelyn. *Yaqui Resistance and Survival: The Struggle for Land and Autonomy, 1821–1910*. Madison: University of Wisconsin Press, 1984.

Jacoby, Karl. *Crimes Against Nature: Squatters, Poachers, Thieves and the Hidden History of American Conservation*. Berkeley: University of California Press, 2001.

Janvier, Thomas A. *The Mexican Guide*. New York: Charles Scribner's Sons, 1886.

Jay, Martin. *Downcast Eyes: The Denigration of Vision in Twentieth-Century French Thought*. Berkeley: University of California Press, 1993.

Jiménez Mora, Adriana. "El proyecto de división territorial de Manuel Orozco y Berra en el Segundo Imperio: Antecedentes, aplicación y problemas." Tesis de Licenciatura en Historia, Facultad de Filosofía y Letras, Universidad Nacional Autónoma de México, 2003.

Johnson, John J. *Latin America in Caricature*. Austin: University of Texas Press, 1980.

Joseph, Gilbert M., and Timothy J. Henderson, eds. *The Mexico Reader: History, Culture, Politics*. Durham: Duke University Press, 2003.

Joseph, Gilbert M., and Daniel Nugent, eds. *Everyday Forms of State Formation: Revolution and the Negotiation of Rule in Modern Mexico*. Durham: Duke University Press, 1994.

———. "Popular Culture and State Formation in Revolutionary Mexico." In Joseph and Nugent, eds., *Everyday Forms of State Formation*.

Justo Alvarez, José. "Departamento de Ingenieros." *Memoria de la Secretaría de Guerra y Marina por Pedro Ogazón*. Mexico City: Tipografía de Gonzalo A. Esteva, 1878.

Kain, Roger J. P., and Elizabeth Baigent. *The Cadastral Map in the Service of the State: A History of Property Mapping*. Chicago: University of Chicago Press, 1992.

Keen, Benjamin. *The Aztec Image in Western Thought*. New Brunswick: Rutgers University Press, 1971.

Kivelson, Valerie. "Cartography, Autocracy and State Powerlessness: The Uses of Maps in Early Modern Russia." *Imago Mundi* 51 (1999), 83–105.

Klor de Alva, José Jorge. "The Postcolonization of the (Latin) American Experience: A Reconsideration of 'Colonialism,' 'Postcolonialism' and 'Mestizaje.'" In *After Colonialism: Imperial Histories and Postcolonial Displacements*, edited by Gyan Prakash Princeton: Princeton University Press, 1995.

Knight, Alan. *The Mexican Revolution*. 2 vols. Lincoln: University of Nebraska Press, 1986.

———. "Racism, Revolution, and Indigenismo: Mexico, 1910–1940." In *The Idea of Race in Latin America, 1870–1940*, edited by Richard Graham. Austin: University of Texas Press, 1990.

Kourí, Emilio. "The Business of the Land: Agrarian Tenure and Enterprise in Papantla, Mexico, 1800–1910." Ph.D. dissertation, Harvard University, 1996.

———. "Economía y comunidad en papantla: Relexiones sobre 'la cuestión de la tierra' en el siglo XIX." In Antonio Escobar y Teresa Rojas, *Lo Agrario en la Historia de México*. Mexico City: CIESAS, 2002.

———. "Interpreting the Expropriation of Indian Pueblo Lands in Porfirian Mexico: The Unexamined Legacies of Andrés Molina Enríquez." *Hispanic American Historical Review* 82, no. 1 (February 2002), 69–117.

Kroeber, Clifton B. *Man, Land and Water: Mexico's Farmlands Irrigation Policies, 1885–1911*. Berkeley: University of California Press, 1983.

Kropotkin, Peter. "What Geography Ought to Be." *The Nineteenth Century* 18 (1885), 940–56.

Kula, Witold. *Measures and Men*. Translated by R. Szreter. Princeton: Princeton University Press, 1986.

Latour, Bruno. *Science in Action: How to Follow Scientists and Engineers through Society*. Cambridge: Harvard University Press, 1987.

Latour, Bruno, and Steve Woolgar. *Laboratory Life: The Social Construction of Scientific Facts*. Beverly Hills: Sage Publications, 1979.

Lauria-Santiago, Aldo. *An Agrarian Republic: Commercial Agriculture and the Politics of Peasant Communities in El Salvador, 1824–1918*. Pittsburgh: University of Pittsburgh Press, 1999.

———. "Land, Community and Revolt in Late-Nineteenth-Century Indian Izalco, El Salvador." *Hispanic American Historical Review* 79, no. 3 (August 1999), 495–534.

Lefebvre, Henri. *The Production of Space*. Translated by Donald Nicholson-Smith. Oxford: Blackwell, 1991.

Lenoir, Timothy. "Inscription Practices and Materialities of Communication." In Lenoir, ed. *Inscribing Science*.

Lenoir, Timothy, ed. *Inscribing Science: Scientific Texts and the Materiality of Communication*. Stanford: Stanford University Press, 1998.

León-Portillo, Miguel, and Carmen Aguilera. *Mapa de México-Tenochtitlan y sus contornos hacia 1550*. Mexico City: Celanese Mexicana, 1986.

Locke, John. *Two Treatises of Government*. Edited by Peter Laslett. Cambridge: Cambridge University Press, 1960.

Lozano Meza, María. "La Sociedad Mexicana de Geografía y Estadística (1833–1867). Un estudio de caso: La estadística." Tesis de Licenciatura en Historia, Facultad de Filosofía y Letras, Universidad Nacional Autónoma de México, 1991.

Maier, Charles S. "Consigning the Twentieth Century to History: Alternative Narratives for the Modern Era." *American Historical Review* 105, no. 3 (June 2000), 807–31.

Mallon, Florencia. *Peasant and Nation: The Making of Postcolonial Mexico and Peru.* Berkeley: University of California Press, 1995.

———. "Reflections on the Ruins: Everyday Forms of State Formation in Nineteenth-Century Mexico." In Joseph and Nugent, eds., *Everyday Forms of State Formation.*

Manero, Vicente. *Documentos interesantes sobre colonización.* Mexico City: Tipografía de la V. e Hijos de Murguia, 1878.

———. "Informe de la Sección Primera." *Memoria presentada al Congreso de la Unión . . . 1877.*

———. "Memoria de la Comisión Geográfico-Exploradora." *El Mensajero— Organo del Partido Liberal Constitucionalista* 1:60 (June 18, 1880).

María Rodríguez, Joaquín. *Apuntes sobre el Cantón de Xalapa, Estado de Veracruz, Mexico.* Xalapa: Imprenta Veracruzana de la Viuda e Hijos de Ruiz, 1895.

———. *Perfiles del suelo Vera Cruzano.* Xalapa-Enríquez: Tipografía del Gobierno del Estado, 1900.

Mason, R. H. *Pictures of Life in Mexico.* London: Smith, Elder, and Co., 1851.

Massey, Doreen. "Places and Their Pasts." *History Workshop Journal* 39 (spring 1995), 182–92.

———. "Politics and Space/Time." *New Left Review* 196 (November-December 1992), 65–84.

———. *Space, Place and Gender.* Oxford: Polity Press, 1994.

Mayer, Brantz. *Mexico: Aztec, Spanish and Republican: A Historical, Geographical, Political, Statistical and Social Account of the Country from the Period of the Invasion by the Spaniards to the Present Time, in Two Volumes.* Hartford: S. Drake, 1851.

Mayer Celis, Leticia. *Entre el infierno de una realidad y el cielo de un imaginario: Estadística y comunidad en el México de la primera mitad del siglo xix.* Mexico City: El Colegio de México, 1999.

McBride, George McCutchen. *The Land Systems of Mexico.* New York: American Geographical Society, 1923.

McClintock, Anne. *Imperial Leather: Race, Gender, and Sexuality in the Colonial Contest.* New York: Routledge, 1995.

McGowan, Gerald L. *Geografía político administrativa de la Reforma: Una visión histórica.* Aguascalientes: INEGI, 1991.

Meinig, D. W. "The Continuous Shaping of America: A Prospectus for Geographers and Historians." *American Historical Review* 83, no. 5 (1978), 1186–1205.

———. *The Shaping of America: A Geographical Perspective on 500 Years of History.* Vol. 2, *Continental America, 1800–1867.* New Haven: Yale University Press, 1993.

Memoria de Hacienda y Crédito Público correspondiente al quincuagísimo cuarto año económica trascurrido del 10. de julio de 1878 al 30 de junio de 1879. Mexico City: Imprenta del Comercio de Dublán y comp., 1880.

Memoria de Hacienda y Crédito Público correspondiente al año fiscal trascurrido de 1 de julio de 1883 a 30 de junio de 1884 presentada por el Secretario de Estado y del despacho de Haciendo y Crédito Público, General Miguel de la Peña, al Congreso de la Unión. Mexico City: Tipografía "La Luz" de Antonio B. de Lara, 1884.

Memoria de la Secretaría de Fomento presentada al Congreso de la Unión por el Secretario de Estado y del despacho del Ramo, Lic. Olegario Molina, corresponde al ejercicio fiscal de 1909–1910. Mexico City: Imprenta y Fototipia de la Secretaría de Fomento, 1910.

Memoria de la Secretaría de Hacienda correspondiente al ejercicio fiscal de 1884 a 1885 presentada al Congreso de la Unión por el Ministro del Ramo, Lic. Manuel Dublán. Mexico City: Imprenta del Gobierno Federal, 1884.

Memoria de la Secretaría de Hacienda correspondiente al año fiscal de 1880 a 1881. Mexico City: Tipografía de Gonzalo A. Esteva, 1881.

Memoria leida por el C. Gobernador del Estado ante la H. Legislatura del mismo, el día 13 de octubre de 1871. Veracruz: Imprenta del "Progreso" de R. Laine, 1871.

Memoria presentada al Congreso de la Unión por el Secretario de Estado y del despacho de Fomento, Colonización, Industria y Comercio, 1877. Mexico City: Imprenta de Francisco Díaz de León, 1877.

Memoria presentada al Congreso de la Union, por el Secretario de Estado y del despacho de Fomento, Colonización, Industria y Comercio de la República Mexicana, General Carlos Pacheco. Corresponde a los años trascurridos de enero de 1883 a junio de 1885. 5 vols. Mexico City: Oficina Tipográfico de la Secretaría de Fomento, 1887.

Memoria presentada al Congreso de la Union, por el Secretario de Estado y del despacho de Fomento, Colonización, Industria y Comercio, corresponde a los años trascurridos de enero de 1883 a junio de 1885. Tomo 6, *Atlas.* Mexico City: Oficina Tipográfico de la Secretaría de Fomento, 1887.

Memoria presentada al Congreso de la Unión por el Secretario de Estado y del despacho de Fomento, Colonización e Industria de la República Mexicana, Ingeniero Manuel Fernández Leal, corresponde a los años transcurridos de 1892 a 1896. Mexico City: Oficina Tipográfica de la Secretaría de Fomento, 1897.

Memoria presentada al Congreso de la Unión por el Secretario de Estado y del despacho de Fomento, Colonización e Industria de la República Mexicana, corresponde a los años transcurridos de 1897 a 1900 y a la gestión administrativa del señor Ingeniero Don Manuel Fernández Leal. Mexico City: Imprenta y Fototipia de la Secretaría de Fomento, 1908.

Memoria presentada al Congreso de la Unión por el Secretario de Estado y del despacho de Fomento, Colonización e Industria de la República Mexicana, corresponde a los años transcurridos de 10. de enero de 1901 al 31 de diciembre de 1904 y a la gestión administrativa de los señores Ing. D. Leandro Fernández Leal y Gral. D. Manuel

González Cosío. Mexico City: Imprenta y Fototipia de la Secretaría de Fomento, 1909.

Memoria presentada al Congreso de la Unión por el Lic. Olegario Molina, Secretario de Estado y del despacho de Fomento, Colonización e Industria de la República Mexicana, corresponde al ejercicio fiscal de 1907–1908. Mexico City: Imprenta y Fototipia de la Secretaría de Fomento, 1910.

Memoria presentada al Congreso de la Unión por el Lic. Olegario Molina, Secretario de Estado y del despacho de Fomento, Colonización e Industria de la República Mexicana, corresponde al ejercicio fiscal de 1908–1909. Mexico City: Imprenta y Fototipia de la Secretaría de Fomento, 1910.

Memoria presentada a la H. Legislatura del Estado Libre y Soberano de Veracruz Llave, el 18 de Septiembre de 1890, por el Gobernador Constitucional C. General Juan Enríquez, comprende el periodo corrido de 1o. de julio de 1888 a 30 de junio de 1890. Jalapa: Imprenta del Gobierno del Estado, 1891.

Memoria que el Secretario de Estado y del despacho de Guerra y Marina, General de División Gerónimo Treviño, presentada al Congreso de la Unión en 31 de mayo de 1881, y comprende del 1o. de diciembre de 1877 a la expresada fecha. 3 vols. Mexico City: Tipografía de Gonzalo A. Esteva, 1881.

Memoria que el Secretario de Estado y del despacho de Guerra y Marina presenta al Congreso de la Unión y que comprende de 1o. de julio de 1886 a 30 de junio de 1890. Mexico City: Tipografía del Cuerpo de Estado Mayor Especial, 1891.

Mendoza Vargas, Héctor. "Historia de la geografía en México: Siglo XIX." Tesis de Licenciatura en Geografía, Facultad de Filosofía y Letras, Universidad Nacional Autónoma de México, 1989.

———. "Los ingenieros geógrafos de México, 1823–1915." Tesis de Maestría en Geografía, Facultad de Filosofía y Letras, Universidad Nacional Autónoma de México, 1993.

———. "Las opciones geográficas al inicio del México independiente." In Mendoza Vargas, ed., *Mexico a través de los mapas*.

———, ed. *Lecturas geográficas mexicanas: Siglo XIX*. Mexico City: UNAM, 1999.

———, ed. *México a través de los mapas*. Mexico City: Plaza y Valdés Editores, 2000.

Mermin, N. David. "Writing Physics." In *Writing and Revising the Disciplines*, edited by Jonathan Monroe. Ithaca: Cornell University Press, 2002.

Meyer, Michael C. *Huerta: A Political Portrait*. Lincoln: University of Nebraska Press, 1972.

Michael, Bernardo. "Separating the Yam from the Boulder: Statemaking, Space, and the Causes of the Anglo-Gorkha War of 1814–1816 (Nepal)." Ph.D. dissertation, University of Hawaii, 2001.

———. "When Soldiers and Statesmen Meet: 'Ethnographic Moments' on the Frontiers of Empire, 1800–1815." Unpublished manuscript in author's possession.

Mignolo, Walter. "Colonial Situations, Geographical Discourses, and Territorial Representations: Toward a Diatopical Understanding of Colonial Semiosis." *Dispositio* 14, nos. 36–38 (1989), 93–140.

———. *The Darker Side of the Renaissance: Literacy, Territoriality, and Colonization.* Ann Arbor: University of Michigan Press, 1995.

Miguel Macías, José. *Xalapa o Jalapa? Artículos publicados en el "Diario Comercial" de Veracruz durante el mes de enero de 1893.* Xalapa-Enríquez: Imprenta del Gobierno del Estado, 1893.

Mitchell, Don. *Cultural Geography: A Critical Introduction.* Oxford: Blackwell, 2000.

Mitchell, Katharyne. "Different Diasporas and the Hype of Hybridity." *Environment and Planning D: Society and Space* 15, no. 5 (1997), 533–53.

Mitchell, Timothy. *Colonising Egypt.* Cambridge: Cambridge University Press, 1988.

———. "The Limits of the State: Beyond Statist Approaches and Their Critics." *American Political Science Review* 85, no. 1 (1991), 77–96.

Molina, Ignacio. "Informe del Jefe del Departamento de Cartografía." *Memoria presentada al Congreso de la Unión . . . enero de 1883 a junio de 1885.*

Molina Enríquez, Andrés. *Los grandes problemas nacionales.* Mexico City: Ediciones Era, 1978 [1909].

Montesinos, José. "Relación del personal de la Secretaría de Guerra y Marina en 31 de Mayo de 1881." *Memoria que el Secretario de Estado y del Despacho de Guerra y Marina . . . 10. de diciembre de 1877 a la expresada fecha.*

Moral, Tomás Ramón del. "Condiciones del trabajo geográfico de la Comisión de Geografía y Estadística del Estado de México, 1827–29." In Mendoza Vargas, ed., *Lecturas geográficas mexicanas: Siglo XIX.*

Mundy, Barbara E. *The Mapping of New Spain: Indigenous Cartography and the Maps of the* Relaciones Geográficas. Chicago: University of Chicago Press, 1996.

Murphy, Arthur D. "To Title or not to Title: Article 27 and Mexico's Urban Ejidos." *Urban Anthropology* 23, nos. 2–3 (summer–fall 1994), 209–32.

Nava Oteo, Guadalupe. "La minería bajo el porfiriato." In Cardoso, ed., *México en el siglo XIX.*

Noriega, Eduardo. "Los progresos de la geografía de Mexico." *Boletín de la Sociedad Mexicana de Geografía y Estadística,* 4a. época, tomo 4 (1897), 264–69.

Nugent, Daniel. *Spent Cartridges of Revolution: An Anthropological History of Namiquipa, Chihuahua.* Chicago: University of Chicago Press, 1993.

Nugent, Daniel, and Ana María Alonso. "Multiple Selective Traditions in Agrarian Reform and Agrarian Struggle: Popular Culture and State Formation in the Ejido of Namiquipa, Chihuahua." In Joseph and Nugent, eds., *Everyday Forms of State Formation.*

Nuijten, Monique. "Changing Legislation and a New Agrarian Bureaucracy: Shifting Arenas of Negotiation." In Zendejas and de Vries, eds., *Rural Transformations Seen from Below.*

———. "Family Property and the Limits of Intervention: The Article 27 Reforms and the PROCEDE Programme in Mexico." *Development and Change* 34, no. 3 (2003), 475–97.

———. *Power, Community, and the State: The Political Anthropology of Organisation in Mexico.* London: Pluto Press, 2003.

———. "Recuerdos de la tierra: Luchas locales e historias fragmentadas." In *Las disputas por el México rural: Transformaciones de prácticas, identidades y proyectos,* vol. 2, *Historias y narrativas,* edited by Sergio Zendejas and Pieter de Vries. Zamora, Michoacán: El Colegio de Michoacán, 1998.

Ober, Frederick A. *Mexican Resources: A Guide to and through Mexico.* Boston: Estes and Lauriat, 1884.

Obeyesekere, Gananath. *The Apotheosis of Captain Cook: European Mythmaking in the Pacific.* Princeton: Princeton University Press, 1995.

O'Brien, Jay, and William Roseberry, eds. *Golden Ages, Dark Ages: Imagining the Past in Anthropology and History.* Berkeley: University of California Press, 1991.

O'Gorman, Edmundo. *Historia de las divisiones territoriales de México.* Mexico City: Editorial Porrua, 1966 [1937].

———. *The Invention of America: An Inquiry into the Historical Nature of the New World and the Meaning of Its History.* Bloomington: Indiana University Press, 1961.

O'Hanlon, Rosalind, and David Washbrook. "After Orientalism: Culture, Criticism and Politics in the Third World." *Comparative Studies in Society and History* 34, no. 1 (1992), 141–67.

———. "Histories in Transition: Approaches to the Study of Colonialism and Culture in India." *History Workshop Journal* 32 (1991), 110–27.

Olavarría y Ferrari, Enrique de. *La Sociedad Mexicana de Geografía y Estadística.* Mexico City: n.p., 1901.

Orlove, Benjamin. "Mapping Reeds and Reading Maps: The Politics of Representation in Lake Titicaca." *American Ethnologist* 18, no. 1 (February 1991), 3–38.

———. "The Ethnography of Maps: The Cultural and Social Contexts of Cartographic Representation in Peru." *Cartographica* 30, no. 1 (1993), 29–46.

Orozco y Berra, Manuel. *Apuntes para la historia de la geografía en México.* Mexico City: Imprenta de Francisco Díaz de León, 1881.

———. *Geografía de las lenguas y carta etnográfica de México.* Mexico City: Imprenta de J. M. Andrade y F. Escalante, 1864.

———. *Historia antigua y de la conquista de México.* Mexico City: Tipografía de G. A. Esteva, 1880.

————. *Materiales para una cartografia mexicana.* Mexico City: Imprenta del Gobierno, 1871.

O'Sullivan, John. "The Texas Question." *United States Democratic Review* 14, no. 70 (April 1844), 423–30.

Ortega y Medina, Juan, and Rosa Camelo, eds. *En busca de un discurso integrador de la nación, 1848–1884.* Mexico City: Universidad Nacional Autónoma de México, 1996.

Otero, Mariano. *Ensayo sobre el verdadero estado de la cuestión social y política que se agita en la República Mexicana.* Mexico City: Impreso por Ignacio Cumplido, 1842.

Overmyer-Velázquez, Mark. "Visions of the Emerald City: Politics, Culture and Alternative Modernities in Oaxaca City, Mexico, 1877–1920." Ph.D. dissertation, Yale University, 2002.

Palacios, Guillermo. "El General de la Nación, el General Agrario." *Boletín del Archivo General Agrario* 8 (October–December 1999), 9–19.

————. "Julio Cuadros Caldas: Una agrarista colombiano en la Revolución Mexicana: Pequeña biografía incidental." In Cuadros Caldas, *Catecismo Agrario.*

Pasquel, Leonardo. *La revolución en el Estado de Veracruz.* 2 vols. Mexico City: Talleres Gráficos de la Nación, 1972.

Peluso, Nancy Lee. *Rich Land, Poor People: Resource Control and Resistance in Java.* Berkeley: University of California Press, 1991.

————. "Whose woods are these? Counter-mapping Forest Territories in Kalimantan, Indonesia." *Antipode* 27, no. 4 (1995), 383–406.

Peñafiel, Antonio. *Nombres geográficos de México: Catálogo alfabético de los nombres de lugar pertenecientes al idioma "Nahuatl," estudio jeroglífico de la matrícula de los tributos del Códice mendocino.* Mexico City: Oficina Tipográfica de la Secretaría de Fomento, 1885.

————. *Nomenclatura geográfica de México: Etimologías de los nombres de lugar correspondientes a los principales idiomas que se hablan en la República.* Mexico City: Oficina Tipográfica de la Secretaría de Fomento, 1897.

Peréz, Matilde, and Victor Ruíz Arrazola. "La solución en la región no será unilateral." *La Jornada,* August 9, 1999, 48.

Pérez Rosales, Laura. "Manuel Orozco y Berra." In Ortega y Medina and Camelo, eds., *En busca de un discurso integrador.*

Pimentel, Francisco. *La economía política aplicada a la propiedad territorial en México.* Mexico City: Imprenta de Ignacio Cumplido, 1866.

Pisa, Rosario. "Popular Response to the Reform of Article 27: State Intervention and Community Resistance in Oaxaca." *Urban Anthropology* 23, nos. 2–3 (summer–fall, 1994), 267–306.

Poblett Miranda, Martha, ed. *Cien viajeros en Veracruz: Crónicas y relatos.* 10 vols. Xalapa-Enríquez: Imprenta del Gobierno del Estado de Veracruz, 1992.

Polanyi, Karl. *The Great Transformation.* New York: Rinehart, 1944.

Poole, Deborah. "Landscape and the Imperial Subject: U.S. Images of the Andes, 1859–1930." In *Close Encounters of Empire: Writing the Cultural History of U.S.-Latin American Relations*, edited by Gilbert M. Joseph, Catherine C. Legrand, and Ricardo D. Salvatore. Durham: Duke University Press, 1998.

Poulantzas, Nicos. *Political Power and Social Classes*. London: Verso Press, 1987.

Prakash, Gyan. "Can the Subaltern Ride? A Reply to O'Hanlon and Washbrook." *Comparative Studies in Society and History* 34, no. 1 (1992), 168–84.

———. "Writing Post-Orientalist Histories of the Third World: Perspectives from Indian Historiography." *Comparative Studies in Society and History* 32, no. 2 (1990), 383–408.

Pratt, Geraldine. "Spatial Metaphors and Speaking Positions." *Environment and Planning D: Society and Space* 10 (1992), 241–44.

Pratt, Mary Louise. *Imperial Eyes: Travel Writing and Transculturation*. New York: Routledge, 1992.

Pred, Allan. *Lost Words and Lost Worlds: Modernity and the Language of Everyday Life in Late Nineteenth-Century Stockholm*. Cambridge: Cambridge University Press, 1990.

———. "Place as Historically Contingent Process: Structuration and the Time-Geography of Becoming Places." *Annals of the Association of American Geographers* 74, no. 2 (1984), 279–97.

Puijol, José F. *Guía del propietario de terrenos*. Mexico City: Imprenta de J. M. Aguilar Ortíz, 1878.

Rama, Angel. *The Lettered City*. Translated by John Charles Chasteen. Durham: Duke University Press, 1996.

Ramírez, Romero. "Proyecto de ley relativo a la conservación de monumentos arqueológicos, agosto 28 de 1862." *Boletín de la Sociedad Mexicana de Geografía y Estadística* 1a. época, tomo 9 (1865), 197–199.

Rappaport, Joanne. *The Politics of Memory: Native Historical Interpretation in the Colombian Andes*. Durham: Duke University Press, 1998.

Rebert, Paula. *La Gran Línea: Mapping the United States—Mexico Boundary, 1849–1857*. Austin: University of Texas Press, 2001.

Reclus, Elisée. *L'Homme et la terre*. Paris: Librairie Universelle, 1905.

Richards, Thomas. *The Imperial Archive: Knowledge and the Fantasy of Empire*. London: Verso Press, 1993.

Richardson, Judith. *Possessions: The History and Uses of Haunting in the Hudson Valley*. Cambridge: Harvard University Press, 2003.

Ricouer, Paul. *Time and Narrative*. Chicago: University of Chicago Press, 1983.

Riguzzi, Paolo. "México próspero: Las dimensiones de la imágen nacional en el porfiriato." *Historias* 20 (1988), 137–58.

Riva Palacio, Vicente, ed. *México a través de los siglos: Historia general y completa del desenvolvimiento social, político, religioso, militar, artístico, científico y literario de México desde la antigüedad más remota hasta la época actual.* 5 vols. Barcelona: Espasa, 1887–1889.

Robinson, Cecil, ed. and trans. *The View from Chapultepec: Mexican Writers on the Mexican-American War.* Tucson: University of Arizona Press, 1989.

Roldán, Mary. *Blood and Fire: La Violencia in Antioquia, Colombia, 1946–1953.* Durham: Duke University Press, 2002.

Romero, José Guadalupe. "Dictamen sobre los inconvenientes de mudar los nombres geográficos de las poblaciones de la República aprobado por la Sociedad." *Boletín de la Sociedad Mexicana de Geografía y Estadística,* 1a. época, tomo 8 (1860), 387–89.

Romero, Matias. *México and the United States: A Study of Subjects Affecting Their Political, Commercial, and Social Relations, Made with a View to their Promotion.* New York and London: G. P. Putnam's Sons, 1898.

Rose, Carol. *Property and Persuasion: Essays on the History, Theory, and Rhetoric of Ownership.* Boulder: Westview Press, 1994.

Roseberry, William. "Hegemony and the Language of Contention." In Joseph and Nugent, eds., *Everyday Forms of State Formation.*

Ross, Kristin. *The Emergence of Social Space: Rimbaud and the Paris Commune.* Minneapolis: University of Minnesota Press, 1988.

Rotman, Brian. "The Technology of Mathematical Persuasion." In Lenoir, ed., *Inscribing Science.*

Ruedas de la Serna, Jorge A. *Los orígenes de la visión paradisiaca de la naturaleza mexicana.* Mexico City: Universidad Nacional Autónoma de México, 1987.

Ruiz Naufal, Víctor Manuel. "La faz del terruño: Planos locales y regionales, siglos XVI–XVIII." In Mendoza Vargas, ed. *México a través de los mapas.*

Sahlins, Marshall. *Stone-Age Economics.* Chicago: Aldine, 1974.

Sahlins, Peter. *Boundaries: The Making of France and Spain in the Pyrenees.* Berkeley: University of California Press, 1989.

Said, Edward. *Culture and Imperialism.* New York: Vintage Books, 1994.

———. *Orientalism.* New York: Vintage Books, 1979.

San Juan Victoria, Carlos, and Salvador Velázquez Ramírez. "La formación del estado y las políticas económicas (1821–1880)." In Ciro Cardoso, ed., *México en el siglo XIX.*

Sánchez Lamego, Miguel A. "Agustín Díaz, ilustre cartógrafo mexicano." *Historia Mexicana* 96 (April–June 1975), 556–65.

Sartorius, Carl. *Mexico: Landscapes and Popular Sketches.* London: Trübner, 1859.

Sayer, Derek. "Everyday Forms of State Formation: Some Dissident Remarks on 'Hegemony.'" In Joseph and Nugent, eds., *Everyday Forms of State Formation.*

Schaffer, Simon. "The Leviathan of Parsonstown: Literary Technology and Scientific Representation." In Lenoir, ed., *Inscribing Science.*

Schama, Simon. *Landscape and Memory.* New York: HarperCollins, 1995.

Schmidt, Alfred. *The Concept of Nature in Marx.* Translated by Ben Fowkes. London: New Left Books, 1971.

Schoultz, Lars. *Beneath the United States: A History of U.S. Policy toward Latin America.* Cambridge: Harvard University Press, 1998.

Schryer, Frans. "Peasants and the Law: A History of Land Tenure and Conflict in the Huasteca." *Journal of Latin American Studies* 18, no. 2 (November 1986), 283–311.

Scott, James C. *The Moral Economy of the Peasant: Rebellion and Subsistence in Southeast Asia.* New Haven: Yale University Press, 1976.

———. *Seeing Like a State: How Certain Schemes to Improve the Human Condition Have Failed.* New Haven: Yale University Press, 1998.

Secretaría de la Reforma Agraria: Perfil agrario del Estado de Veracruz: Delegación Xalapa, 1915–1979. Unpublished manuscript in the possession of Ingeniero Héctor Rivadeneyra.

Seed, Patricia. *Ceremonies of Possession in Europe's Conquest of the New World, 1492–1640.* Cambridge: Cambridge University Press, 1995.

Sellers, Charles. *The Market Revolution: Jacksonian America, 1815–1846.* Oxford: Oxford University Press, 1991.

Shapin, Steven, and Simon Schaffer. *Leviathan and the Air Pump: Hobbes, Boyle and the Experimental Life.* Princeton: Princeton University Press, 1985.

Shepard, A. K. *The Land of the Aztecs, or Two Years in Mexico.* Albany: Weed, Parsons, and Co., 1859.

Siemens, Alfred. *Between the Summit and the Sea: Central Veracruz in the Nineteenth Century.* Vancouver: University of British Columbia Press, 1990.

Simpson, Eyler N. *The Ejido: Mexico's Way Out.* Chapel Hill: University of North Carolina Press, 1937.

Smith, Neil. *Uneven Development: Nature, Capital and the Production of Space.* Oxford: Blackwell, 1984.

Smith, Neil, and Cindi Katz. "Grounding Metaphor: Towards a Spatialized Politics." In *Place and the Politics of Identity,* edited by Michael Keith and Steve Pile. New York: Routledge, 1993.

Soja, Edward W. *Postmodern Geographies: The Reassertion of Space in Critical Social Theory.* London: Verso, 1989.

Somers, Margaret. "The Narrative Constitution of Identity: A Relational and Network Approach." *Theory and Society* 23 (1994), 605–49.

Soothill, William Edward. *The Analects of Confucius.* New York: Paragon Book Reprint Corp., 1968.

Stephen, Lynn. "Accommodation and Resistance: Ejidatario, Ejidataria, and Official Views of Ejido Reform." *Urban Anthropology* 23, nos. 2–3 (summer–fall, 1994), 233–66.

Suárez Cortez, Blanca Estela, ed. *Historia de los usos del agua en México: Oligarquías, empresas y ayuntamientos (1840–1940).* Mexico City: CIESAS-IMTA, 1998.

Tamayo P. de Ham, Luz María Oralia. *La geografía: Arma científica para la defensa del territorio.* Mexico City: Plaza y Valdés Editores, 2001.

Taylor, William. *Landlord and Peasant in Colonial Oaxaca.* Stanford: Stanford University Press, 1972.

Tenenbaum, Barbara. "Streetwise History: The Paseo de la Reforma and the Porfirian State, 1876–1910." In Beezley, Martin, and French, eds., *Rituals of Rule, Rituals of Resistance.*

Tenorio-Trillo, Mauricio. *Mexico at the World's Fairs: Crafting a Modern Nation.* Berkeley: University of California Press, 1996.

———. "1910 Mexico City: Space and Nation in the City of the Centenario." *Journal of Latin American Studies* 28, no. 1 (February 1996), 75–104.

Thompson, E. P. "Custom, Law and Common Right." In Thompson, *Customs in Common.*

———. *Customs in Common: Studies in Traditional Popular Culture.* New York: New Press, 1993.

———. "The Moral Economy of the English Crowd in the Eighteenth Century." In Thompson, *Customs in Common.*

———. "The Moral Economy Reviewed." In Thompson, *Customs in Common.*

Thomson, Guy. "'La Republique au Village' in Spain and Mexico, 1848–1888." In *Nation Building in Nineteenth-Century Latin America: Dilemmas and Conflicts,* edited by Hans-Joachim Konig and Marriane Wiesebron. Leiden: Leiden University, CNWS Publications, 1998.

Thongchai Winichakul. *Siam Mapped: The History of the Geo-Body of a Nation.* Honolulu: University of Hawaii Press, 1994.

Toulmin, Stephen. *Cosmopolis: The Hidden Agenda of Modernity.* New York: Free Press, 1990.

Trabulse, Elias. *Cartografía mexicana: Tesoros de la nación siglos XVI a XIX.* Mexico City: Archivo General de la Nación, 1983.

Trachtenberg, Alan. *Reading American Photographs: Images as History, Matthew Brady to Walker Evans.* New York: Hill and Wang, 1989.

Tratado de paz, amistad y límites entre la república Mexicana y los estados-unidos de norte América. Veracruz: Imprenta de Comercio, 1848.

Trens, Manuel B. *Historia de Veracruz.* 6 vols. Mexico City: S. Turanzas de Valle, 1950.

Trouillot, Michel-Rolph. *Silencing the Past: Power and the Production of History.* Boston: Beacon Press, 1995.

Tuan, Yi-Fu. *Space and Place: The Perspective of Experience.* Minneapolis: University of Minnesota Press, 1977.

Turnbull, David. *Maps Are Territories: Science Is an Atlas.* Chicago: University of Chicago Press, 1993.

Tutino, John. "Agrarian Social Change and Peasant Rebellion in Nineteenth-Century Mexico: The Example of Chalco." In *Riot, Rebellion, and Revolution:*

Rural Social Conflict in Mexico, edited by Friedrich Katz. Princeton: Princeton University Press, 1988.

———. *From Insurrection to Revolution in Mexico: Social Bases of Agrarian Violence, 1750–1940*. Princeton: Princeton University Press, 1986.

Tylor, Edward B. *Anahuac: or, Mexico and the Mexicans, Ancient and Modern*. London: Longmans, Green, Reader and Dyer, 1861.

Van Young, Eric. "Conflict and Solidarity in Indian Village Life: The Guadalajara Region in the Late Colonial Period." *Hispanic American Historical Review* 64, no. 1 (1984), 55–79.

———. *Hacienda and Market in Eighteenth-Century Mexico: The Rural Economy of the Guadalajara Region, 1675–1820*. Berkeley: University of California Press, 1981.

———. "Paisaje de ensueño con figuras y vallados: Disputa y discurso cultural en el campo mexicano de fines de la Colonia." In *Paisajes rebeldes: Una larga noche de rebelión indígena*, edited by Jane-Dale Lloyd and Laura Pérez Rosales. Mexico City: Universidad Iberoamericana, 1995.

Van Young, Eric, ed. *Mexico's Regions: Comparative History and Development*. La Jolla: Center for U.S.-Mexican Studies, 1992.

Vaughan, Mary Kay. *Cultural Politics in Revolution: Teachers, Peasants and Schools in Mexico, 1930–1940*. Tucson: University of Arizona Press, 1997.

Velasco Toro, José. "Indigenismo y rebelión totonaca de Papantla, 1885–1896." *América Indígena* 39, no. 1 (1979), 81–105.

Villareal Muñoz, Antonio. *Restitución y dotación de ejidos: Codificación de leyes, decretos, y circulares expedidas en materia agraria*. Mexico City: Comisión Nacional Agraria, 1921.

Villoro, Luis. *Los grandes momentos del indigenismo en México*. Mexico City: Secretaría de Educación Pública, 1987.

Wade, Peter. *Blackness and Race Mixture: The Dynamics of Racial Identity in Colombia*. Baltimore: Johns Hopkins University Press, 1992.

Wells, Allen, and Gilbert M. Joseph. *Summer of Discontent, Seasons of Upheaval: Elite Politics and Rural Insurgency in Yucatan, 1876–1915*. Stanford: Stanford University Press, 1996.

White, Hayden. *The Content of the Form: Narrative Discourse and Historical Representation*. Baltimore: Johns Hopkins University Press, 1987.

———. *Metahistory: The Historical Imagination in Nineteenth-Century Europe*. Baltimore: Johns Hopkins University Press, 1973.

Widdifield, Stacie G. *The Embodiment of the National in Late Nineteenth-Century Mexican Painting*. Tucson: University of Arizona Press, 1996.

Williams, Raymond. *The Country and the City*. Oxford: Oxford University Press, 1973.

———. *Marxism and Literature*. Oxford: Oxford University Press, 1977.

Wolfe, Patrick. "History and Imperialism: A Century of Theory, from Marx to

Postcolonialism." *American Historical Review* 102, no. 2 (April 1997), 388–420.

Womack Jr., John. "The Mexican Revolution: 1910–1920." In *Mexico since Independence*, edited by Leslie Bethell. Cambridge: Cambridge University Press, 1991.

———. *Zapata and the Mexican Revolution*. New York: Vintage Books, 1968.

Wood, Denis. *The Power of Maps*. New York: Guilford Press, 1992.

Von Humboldt, Alexander. *Political Essay on the Kingdom of New Spain*. Edited by Mary Maples Dunn. Norman: University of Oklahoma Press, 1972.

Zendejas, Sergio. "Appropriating Governmental Reforms: The Ejido as an Arena of Confrontation and Negotiation." In Zendajes and de Vries, eds., *Rural Transformations Seen from Below*.

Zendejas, Sergio, and Pieter de Vries, eds. *Rural Transformations Seen from Below: Regional Perspectives from Western Mexico*. La Jolla: Center for U.S.-Mexican Studies, 1997.

Zoraida Vázquez, Josefina, and Lorenzo Meyer. *México frente a Estados Unidos: Un ensayo histórico, 1776–1980*. Mexico City: El Colegio de México, 1982.

Zuleta, María Cecelia. "La invención de una agricultura próspera: Itinerarios del fomento agrícola entre el porfiriato y la revolución, 1876–1915." Tesis de doctorado en Historia, El Colegio de México, 2000.

INDEX

Italicized page numbers refer to figures in the text.

Raymond B. Craib is an assistant professor
of history at Cornell University.

Library of Congress Cataloging-in-Publication Data
Craib, Raymond B.
Cartographic Mexico : a history of state fixations and
fugitive landscapes / Raymond B. Craib.
p. cm. — (Latin America otherwise)
Includes bibliographical references and index.
ISBN 0-8223-3405-4 (cloth : alk. paper)
ISBN 0-8223-3416-X (pbk. : alk. paper)
1. Political geography. 2. Cartography—Mexico—History.
3. Geographical perception—Political aspects—Mexico.
4. Land tenure—Mexico—History. 5. Mexico—Historical
geography. I. Title. II. Series.
F1228.9.C73 2004
911'.72—dc22   2004009140